The Self-Marginalization of Wilhelm Stekel

Freudian Circles Inside and Out

PATH IN PSYCHOLOGY

Published in Cooperation with Publications for the Advancement of Theory and History in Psychology (PATH)

CHOICES FOR LIVING: Coping with Fear and Dying
Thoman S. Langner

COGNITIVE SCIENCE AND GENETIC EPISTEMOLOGY: A Case Study of Understanding
David Leiser and Christiana Gillièron

A CRITICAL PSYCHOLOGY: Interpretation of the Personal World
Edmund V. Sullivan

CRITICAL THEORIES OF PSYCHOLOGICAL DEVELOPMENT
Edited by John M. Broughton

CULTURAL PSYCHOLOGY AND QUALITATIVE METHODOLOGY: Theoretical and Empirical Considerations
Carl Ratner

CULTURAL PSYCHOLOGY: Theory and Method
Carl Ratner

DEVELOPMENTAL APPROACHES TO THE SELF
Edited by Benjamin Lee and Gil G. Noam

EDUCATION, ARTS, AND MORALITY: Creative Journeys
Edited by Doris B. Wallace

FRANTZ FANON AND THE PSYCHOLOGY OF OPPRESSION
Hussein Abdilahi Bulhan

HEINZ WERNER AND DEVELOPMENTAL SCIENCE
Edited by Jaan Valsiner

HUMANISTIC PSYCHOLOGY: Concepts and Criticisms
Edited by Joseph R. Royce and Leendert P. Mos

THE LIFE CYCLE OF PSYCHOLOGICAL IDEAS
Edited by Thomas C. Dalton and Rand B. Evans

MANUFACTURING SOCIAL DISTRESS: Psychopathy in Everyday Life
Robert W. Rieber

THE MASS PSYCHOLOGY OF ETHNONATIONALISM
Dusan Kecmanovic

THE PROCESS APPROACH TO PERSONALITY: Perceptgenesis and Kindred Approaches in Focus
Gudmund J. W. Smith

REGIONAL IDENTITY AND BEHAVIOR
Max Sugar

SELF AND IDENTITY IN MODERN PSYCHOLOGY AND INDIAN THOUGHT
Anand C. Paranjpe

WILHELM WUNDT IN HISTORY: The Making of a Scientific Psychology
Edited by Robert W. Rieber and David K. Robinson

The Self-Marginalization of Wilhelm Stekel

Freudian Circles Inside and Out

Jaap Bos

Utrecht University
Utrecht, The Netherlands

Leendert Groenendijk

Vrije Universiteit
Amsterdam, The Netherlands

With contributions by

Johan Sturm

Vrije Universiteit
Amsterdam, the Netherlands

Paul Roazen

York University
Toronto, Canada

Springer

Jaap Bos
Department of Interdisciplinary
Social Science (ASW)
Utrecht University
PO Box 80.140
3508 TC, Utrecht
The Netherlands
j.c.bos@fss.uu.nl

Leendert Groenendijk
Department of Psychology and
Educational Sciences
Vrije Universiteit
1 Van der Boechorststraat
1081 BT, Amsterdam
The Netherlands
lf.groenendijk@psy.vu.nl

Cover drawing by Gertrude Zuckerkandl

ISSN: 1574-048X
ISBN-10: 1-4614-9802-3
ISBN-13: 978-1-4614-9802-5
ISBN-13: 978-0-387-32700-6 (eBook)

Softcover re-print of the Hardcover 1st edition 2007

9 8 7 6 5 4 3 2 1

springer.com

FOREWORD

Like in philosophy, in psychoanalysis the history of the discipline is an essential component of its subject matter, body of observations, and even method. The present work by Bos and Groenendijk about Stekel is a most welcome contribution to scholarly historical research in psychoanalysis. Wilhelm Stekel is today all but forgotten, but in his time he was one of the most prolific psychoanalytic writers. Volume IV alone of Grinstein's *The Index of Psychoanalytic Writings* lists 214 articles and books and 151 abstracts. Stekel achieved considerable fame in Vienna and even more so in the United States, where his books on sexual aberrations were widely read. Among clinicians in the United States, E.A. Gutheil was his most devoted champion. Personally I have had considerable liking for Stekel's ideas, and especially for his notion of the polyphony of thought. Patients come for treatment because of fears and anxieties regarding relationships with other people. But they all are also immensely afraid of their own thoughts. Stekel was unique in understanding such anxieties and analyzed them from many angles. He was also unusually frank about his own sexuality and courageous with dealing with the subject of sexuality in others. I learned an interesting detail about the history of my first psychoanalytic teacher in the early 1960s, the Jewish-Hungarian analyst Sandor Feldman. Apparently Feldman was once threatened with ostracism by the International Psychoanalytic Association because of his adherence to Stekel.

In spite of Stekel's pioneering role in the early days of the psychoanalytic movement, Freud developed an intense dislike toward the man even as he praised some of his contributions. As a result Wilhelm Stekel was marginalized already prior to his expulsion from the psychoanalytic movement in 1912. The authors astutely analyze the social phenomenon of marginalization, the role of Stekel himself in shaping this fate of his, and the behavior of Freud and others in the movement.

Psychoanalysis as a method has some claim of being unitary, as for example in the technique of free association and dream interpretation; but there has been a decline of interest in these areas in the last decades. Also, from its inception, as a body of observation and clinical practice it was preeminently concerned with the many faces of sexuality. Sexuality, too, has fallen by the wayside, at first overshadowed by ego psychology, beginning in the 1920s, and more recently displaced by self-psychology. In connection with sexuality, the authors resurrect a historical quarrel between Freud and Stekel about the alleged harmfulness of masturbation. It turns out that Freud acted out a major prejudice against masturbation, in the manner of medical men and educators of the 19th century whose books and pamphlets filled libraries by the hundreds.

However, the height of debate, dissension, and destruction of careers and lives was reached in the realm of theories. Early on psychoanalysis, a clinical discipline has often, and rather glibly, been attacked as a sect, a cult, a religion. In some ways this analogy is right: the very idea that there are orthodox analysts or an orthodox psychoanalysis, whereas there are orthodox physicists or orthodox physics, supports the contention that in defending certain theories against dissenters Freud acted as a pope persecuting heretics. To use another religious analogy, Freud was practicing the old Jewish herem, or excommunication of heretics, as happened to Spinoza with Amsterdam Jews, or the Holy Inquisition, who burned Giordano Bruno at the stake and nearly condemned Galileo to the same fate.

The history of ideas and the psychoanalytic movement has been studied from two divergent perspectives: the hagiographic, or the official story, e.g., by Ernest Jones or Peter Gay; and the iconoclastic, or revisionist, e.g., Phyllis Grosskurth. Clearly, these two perspectives reflect the orthodoxy bequeathed upon the followers by Freud himself or the keepers of the faith, the famous "Secret Committee." One is reminded of the break-up with Adler and Jung over issues of doctrine as well as the acrimony with which Freud attacked Wilhelm Reich for dissenting from his theory of the death instinct and his subsequent excommunication. As often happens to dissenters, both Wilhelms were not only branded as deviant but also as mentally disturbed.

It goes without saying that other than doctrine, as in the case of the Catholics and the Protestants, the inclusions in and exclusions from of people in the various psychoanalytic societies were always a matter of psychoanalytic politics. But politics has been the blind spot, nay, has been treated with denial, by Freud and the epigones as well. The authors address this issue, too. In this connection I would like to pay a special tribute to the pioneering work of the historian Paul Roazen, co-author of one of the chapters, whose recent untimely death saddened many. It should also be noted that politics of exclusion have been at the heart of warfare among analysts of different persuasions both in Europe and in the United States: Freudians against Jungians; Melanie Klein against Anna Freud; true Freudians vs. neo-Freudians, such as Harald Schultz-Hencke and Karen Horney; the question of medical psychoanalysis and the decades of exclusion of lay psychoanalysis, as happened in the United States.

Studying the lives of prominent dissidents, such as Reich, Ferenczi, and Stekel, who, in turn viewed themselves as true Freudians, is both a fascinating and an indispensable part of this history that is still relevant to us today, to the question where do we come from and where are we going. Illuminating in this context is the comparison the authors provide of two histories of the psychoanalytic movement: Freud and Stekel. And so is the volume in its entirety.

Zvi Lothane, M. D.
Author of In Defense of Schreber/
Soul Murder and Psychiatry
Associate Professor of Psychiatry
Mount Sinai School of Medicine
Member IPA and APsaA

PREFACE

This book became a book when it was half-written already. We had been working on the history of psychoanalysis in general, and on Stekel in particular, independent of each other for some time before we met through a mutual contact. Having discovered that we shared a common interest, we decided to write a paper together. The result of this initial collaboration was published in *The International Journal of Psychoanalysis* as "The art of imitation: Wilhelm Stekel's *Lehrjahre*", which now reappears as Chapter 5 in this book. The focus in this paper is on reading Stekel's and Freud's work in a "dialogical way". The Stekel–Freud relationship was an ambiguous one, we believe, and some of their writings read both as a (covert) reflection of as well as an instrument in this relationship. In trying to lay bare a "conversation between the lines" we focussed on Stekel's formative years and commented on the function of autobiographical narratives in their respective works as one of the mechanisms that produces psychoanalytic discourse.

With our paper on Stekel's formative years published in the summer of 2004 we felt that the idea of reading psychoanalysis dialogically was not yet exhausted, and so it only seemed natural to continue our collaboration. We quickly established that the overarching theme should be the problem of marginalization and self-marginalization, and that one way to understand Stekel's position is to look at it from the perspective of "constructive" or "positive marginality". Accordingly, we rearranged and partly rewrote a number of papers that we had written earlier, which now make up the main gist of this book. While we left the argumentation of each paper largely intact, we smoothened stylistic issues to fit them into a coherent whole, even if some repetition was unavoidable at times.

Thus, while this book discusses in detail the work and influence of Wilhelm Stekel, and his ambiguous relationship to Freud, it is not so much a biography of a "forgotten psychoanalyst", but rather a study in the dimensions of marginality as displayed in the work of one such "marginal author". The various chapters focus on dialogical, interpersonal, social and psychological dimensions of marginality and self-marginality.

The main argument developed in this book is that marginalization had had beneficial aspects to both Stekel and Freud. It gave Stekel room for what could be called *dissentient originality*, while it allowed Freud to use Stekel's dissentience as a counter position against which he would build his arguments. It is in this sense that the Freud–Stekel relationship was formative for many of Freud's relations, both with his own followers as well as with his adversaries.

Divergent opinions between Freud and Stekel with respect to theoretical, clinical and personal matters are discussed in the first seven chapters. The first chapter offers a theoretical exploration of the problem of marginalization and gives a short resumé of the life and work of Wilhelm Stekel, and his struggles with Freud. The ambiguous relationship between Freud and Stekel is further discussed in the second, fourth and seventh chapter, which have an alternating focus on their divergent views on psychoanalytic training, lay-analysis and auto-analysis, respectively. The third chapter takes issue with an important clinical problem on which Freud and Stekel had a long-lasting feud, namely that of the importance of masturbation in the aetiology of neurosis. Chapter 4 discusses the influence of Stekel in the area of mental hygiene and education. The sixth chapter reviews the issue of historiography written from the point of view of self-marginalization, and addresses the problem that the history of psychoanalysis still is, in many ways, determined by Freud's viewpoints. Chapter 7 gives a review of the interpersonal dynamics that can be found in the hitherto unpublished Freud–Stekel correspondence. The last two chapters in this book include the translation of a little-known original piece on Stekel's role in psychoanalytic history, written by Stekel himself in 1926, and the English translation of his correspondence with Freud.

This book would not have been written – or rather not have come about – without the help of many people, some of whom we would like to thank here. First of all we wish to express gratitude to Bob Rieber, to whom credit should go of having suggested that this book be written in the first place. Standing on a hill top overlooking Salzburg, where a century earlier Freud and Stekel once wandered, he encouraged the plan and offered to help find a publisher. We are proud that our book now appears as one of the volumes in his Paths series.

Also, we wish to thank Paul Roazen who volunteered to sort out the Stekel correspondence from the Library of Congress in Washington and helped prepare an introductory chapter to this correspondence. Sadly, Paul died not knowing that his contribution to this book would be printed; his encouragement and confidence in our work, however, enthused us and he will be greatly missed.

Thank you, furthermore Johan Sturm, co-author of the chapter on Stekel's influence on education, for your many valuable suggestions, and Zvi Lothane, for your preface. A warm thank you to Petteri Pietikainen, Dave Park, John Forrester, Michael Molnar, Ben Spiecker, Michael Schröter, Michael Giefer, Mike Sokal, Francis Clark-Lowes, Dave Lee, and Ernst Falzeder for your helpful suggestions and commentaries on parts of this book. We are grateful to the editors and publishers of the *International Journal of Psychoanalysis*, *Psychoanalysis and History*, *History of Psychology*, the *Journal of Psychology and Human Sexuality* and *Paedagogica Historica*, for permission to use papers previously published therein, and to the people at Springer (in particular Sharon Panulla) for their support.

January 2006
Jaap Bos and Leendert Groenendijk

CONTENTS

Chapter 1

MARGINALIZATION THROUGH PSYCHOANALYSIS
An Introduction

Jaap Bos and Leendert Groenendijk

Marginality is becoming universal
Michel de Certeau, 1988, p. xvii.

Prologue

What significance does marginality have in science, if any? Why are *some* scientific practices marginalized while others become dominant, and yes, who *does* the marginalization? Why focus on marginality if there are still so many other interesting topics to explore?

These are some questions we had in mind when we began to explore the position of one particular marginal—Wilhelm Stekel—within a field—psychoanalysis—that was at one time marginalized by other practices and at another a marginalizing practice itself. We soon discovered that what at first sight seemed self-evident, was perhaps not so clear after all. If Stekel was marginalized because his work really did not fit in the corpus of analytical texts, then why was he in psychoanalytic circles for so long? Or, as it is also claimed, if Stekel went "totally astray" after his break with Freud, then in what ways did he digress, and why was it necessary to negate these digressions so strongly? In other words: where or when does marginality emerge (see also Timms, 1992, p. 60)?

Let us begin with a common sense notion of this concept: marginality as the borders of a scientific enterprise. That is, all those activities that are not central to the core business of science. This also includes *people,* the "marginals," whose works are at best not completely nonsensical but also not worth bothering with, or, at worst, must be ignored at all costs. So, marginality can be understood in everyday language to be everything that falls outside the domain of what is considered "interesting" or "pivotal." There is a sharp distinction between the central and the marginal and the two are not to be confused. One should always have only a marginal interest in marginality.

We confess that we have a more than marginal interest in marginality. But we also believe that the above definition includes an illusive notion of marginality. We are going to argue that, paradoxically, marginality is not a marginal phenomenon at all: it is at the heart of all scientific activity. To illustrate this point we turn to what is perhaps *the* most important marginal relationship in psychoanalysis: the one between Sigmund Freud and Wilhelm Stekel, one of his earliest followers who was marginalized in a particularly brutal manner. He was literally kicked out of the "movement," and was never permitted to reenter the psychoanalytic community again in spite of several attempts to reconcile with the master. We believe that the conditions that constituted this relationship were, in a sense, prototypical for future relationships to develop in psychoanalysis. We shall return to this problem at the end of this chapter.

Let us be straightforward then: this book is *not* a biography of Wilhelm Stekel, nor is it, strictly speaking, a collection of historical studies *on* Stekel. Despite the fact that this book includes the complete hitherto unpublished correspondence between Freud and Stekel, and contains an original piece written by Stekel, we do not intend to bring "new facts" to light. However, the studies collected in this volume do reveal new *insights* into the Freud–Stekel relationship—one of the unsolved riddles in the early history of psychoanalysis. Even so, our book is primarily aimed at understanding the work and influence of Stekel *as a marginal*. Briefly, this is a study in the dimensions of marginality, and our aim is to understand this phenomenon by exploring the various dimensions of marginality in the life and work of one marginal in detail. We focus on dialogical, interpersonal, social, and psychological dimensions in the various chapters in this book, but each chapter reads as a separate study that deals with separate issues.

In the remainder of this introductory chapter we shall first outline the problem of marginality somewhat further by positioning it in a theoretical framework. Secondly, we sketch a brief biographical picture of Stekel's life and work, in particular with respect to his problematic relationship with Freud. We then attempt to apply the model of marginalization to the position of Stekel, and incorporate here a few short notions on some of the most important discussions between Freud and Stekel in anticipation of the various issues that are explored in this book. We conclude with a brief discussion of the main characteristics of the prototypical position of a marginal in general.

Marginality as Positioning[1]

Marginality: a classical sociological concept. True, but also a concept that is used in many different ways; so many, in fact, that the question was posed about whether it would not be better for us to dispose of it altogether (Del Pilar & Udasco, 2004). It is, first of all, a concept to describe a *condition* that seems to befall individuals or even groups of people, either by accident or deliberately. As a condition, as a state of being, marginality is understood to be an extremely simple phenomenon: it is a moving away from the center (rich, powerful) to the margins (poor, powerless), assuming that there is a spatial link between the marginal and the central; that the two are physically alienated from each other by time and social forces.

Secondly, marginality is understood to be a *process*. In this definition, the focus is on the traffic between the center and the border—notions of inclusion and exclusion, upward and downward mobility, social integration and segregation are but a few concepts that denote these movements and actions. Thus Bourdieu (1988, 1992), who was particularly interested in the problem of how and through which mechanisms ideas can penetrate into the heart of the business and gain status, called the traffic from the margins to the center "consecration." When marginal ideas become consecrated they may change dominant opinion, but in themselves these ideas also change during this process because they now acquire a new status, and with that—a new meaning.

Yet another way of approaching the question of what it takes to become a marginal is by differentiating between psychological, social as well as institutional mechanisms operative in it. Stonequist (1937), who was one of the first to systematically explore marginality, described the lifecycle of marginal groups as essentially learning to give up one's identity, while Kerckhoff and McCormick (1955) define marginalization in terms of psychological disfunctioning, describing the "marginal man" as socially isolated, an apathetic person who suffers from mood swings, is unable to act decisively, has serious doubts about his social place. In the works of Douglas (1986, 1999) the focus is more on social and societal mechanisms, such as rites of passage, that mark the transgression from the border to the center.

Surely these models are designed in order to better understand the marginal, so as to improve his living conditions or facilitate his transition from the margin to the center. However, even the most idealistic social scientist admits that there is and indeed must be a difference between the marginal and the central, even if this difference is built on a completely arbitrary

[1] This section is partly based on J. Bos, D. Park, & P. Pietikainen (2005). Strategic self-marginalization: The case of psychoanalysis. *Journal of the History of the Behavioral Sciences, 41*, 207–224.

norm. "Between the last person to pass and the first person to fail, the competitive examination creates a difference of all or nothing that can last a lifetime. The former will graduate from an elite institution [...] while the latter will become a nobody" (Bourdieu, 1992, p. 120).

While we are interested in the nobodies, the ones that do *not* pass their exams, we do not want to address the question of why it is that they fail. Instead we would like to draw attention to something that is not accounted for in many of these models of marginality, yet that is clearly recognized in Bourdieu's metaphor: the fact that there is a constant *dialogue* between the marginalizer and the marginalized. In fact, they are not at all separated by time, space, social inhibitions, or laws: The marginalizer and marginalized are bound by a common understanding of what mainstream practices are.

Marginalization is a process that takes place in full consent with the marginalized; in fact, he often actively contributes to it, i.e., to his own marginalization: knowingly and willingly, because the marginal position is a position in its own right, not because he is a victim of it. Nor does marginalization need result in one's expulsion from a group or a society: marginalization, as we want to understand it, is not in the first place about exclusion, but about positioning. The marginal position is a position in its own right, and it obeys its own logic.

Foucault (1973) was one of the first to insist on this, when he gave voice to the "logic of unreason" (madness, crime, sexual deviance, etc.). Ever since it has become easier to see that marginalization can result in a position that has *positive* values, which allow the person to escape from the narrow confinement of the dominant position ascribed to him by social convention. The marginal deliberately situates himself between two cultures or groups, rather than in the center of one. Lumsden (1984) called this the in-between location or the "liminal position." Similarly, Bennett (1997) argued that marginalization can be constructive, rather than destructive or detrimental to one's position, arguing that it allows a person to develop his identity more strongly. And recently, McLauglin (2001) called the marginal position of Erich Fromm "optimal" because it allowed him to carve out a space in the public domain on account of his recalcitrant reading of Freud. McLaughlin (2001, p. 273) argues that optimal marginal intellectuals have access to the creative core of an intellectual tradition, but are not bound by institutional restrictions. They are therefore in an ideal position to transfer novel ideas from the margin to the core.

All these authors undercut the idea that the marginal position can only be understood as the inadequate or lesser version of the dominant position (the marginal as the dominant manqué). In our conception of it, marginalization is an ongoing *dialogic* process, a constituent factor in the wider development of science. The central and the marginal positions are therefore not mutually exclusive, but share a common interest. The position of the marginal contains the *raison d'être* of the dominant, but also the conditions and terms under which the latter are built. Following Bakhtin's notion of dialogism (1981), which emphasizes the continuous communicative acts between self and other in a never ending attempt to position oneself, we find it useful to look at marginalization as a necessary element in a constant redefinition of what is "normal science" that takes place in full consent with the one who is being marginalized. Therefore, the road that leads to marginalization needs to be studied at an interpersonal level as well as a social-discursive level, because it implies forms of negotiation, social positioning, adaptation to and retraction from common viewpoints, and so on. After all, what is perceived to be "dominant" at one moment is the result of a hegemonic struggle, not an objective condition.

Before we try to illustrate these points, we briefly differentiate marginalization from processes that are akin to it, such as repression, domination, stigmatization, and other forms of systematic exclusion on the one hand, and disciplining and other forms of patronizing on the other. What all these processes have in common is a dynamic based on differences in power or status. Stigmatization, for example, is about exercising power, and more in particular about one party defining another party's identity *in terms* of exclusion. This means that in the dominant discourse, there is no place for those who are stigmatized: "identity norms breed deviations as well as conformance" says Goffman (1963, p. 129). Disciplining, on the other hand, is more or less the negative of this: it is about defining the identity of one party in terms of *inclusion*. Those who are being disciplined are going through a process of identity transformation that continues until he "fits in." But it still presupposes a dominant discourse in which one's position is not accepted or recognized as such.

While marginalization is also based on differences in power and status and may also incorporate elements of inclusion and exclusion, it is different from these concepts in the sense that it presupposes active cooperation of two parties. In contradistinction to stigmatization or disciplining, the position of the marginal is acknowledged and in fact functional in the dominant discourse.

In our book we explore the Freud–Stekel relationship from the perspective of "constructive marginality," arguing that Stekel's marginalization had had beneficial aspects for both Stekel and Freud. It gave Stekel room for what could perhaps be called *dissentient originality*: originality that is based on the fact that it contradicts dominant viewpoints. Our approach to marginalization in psychoanalysis is different from that of Pohlen and Bautz-Holzherr (1995, pp. 155–170), who, for example, argue that Freud and his followers were anxious to keep the psychoanalytic canon free from alien influences, and were keen to reprimand and ultimately dispel dissentient voices. The purity of psychoanalytic knowledge was, it is true, always at the center of Freud's attention, but the construction of purity also works to the advantage of the "dissident" who wants to move the limits of a practice because it forces the proponent of purity to expose the weaknesses in the system.

Of course it is true that either way, the marginal comes off worst. When Stekel wrote in his 1926 account of the analytical movement (a translation of which is included in this book) that Freud himself said he always felt the need to have an enemy around, and that he would build one whenever he did not have one at hand, he wrongly believed to have fallen prey to an irrational, yes, evil oppressive mechanism. But he was right about the fact that the authoritative voice of Freud needed an antithetical position before it could be articulated at all.

Stekel enthusiastically cooperated with Freud in what could be called a symbiotic or antagonistic relationship and was driven out of the psychoanalytic community when he began to question the fundamental inequality of their respective roles. It seems to us that this dialectic of antagonism constituted the basis of many psychoanalytic relationships with Freud. We think of Adler, Groddeck, Horney, Jung, Klein, Pfister, Rank, Reich, and others.

In many of these relationships a similar logic developed as in the Freud–Stekel relationship, focusing around common themes such as priority, originality, acknowledgment, and loyalty. In a relationship of marginality, the marginal always struggles with the problem of how to contribute something to the doxa without, at the same time, undermining his own position. Conversely, the question for a dominant is how to use marginal ideas without losing authority. Thus in the case of Groddeck, who insisted on being a marginal even *before* he became a follower of Freud, we find that his main contribution to psychoanalysis, *das Es*, was never acknowledged as his property (his *Es* was even differentiated from Freud's in English by way of a different translation: the *id* (Freud) and the *it* (Groddeck); see Bos, 1992). Similarly, when Rank developed an entire book from a footnote by Freud (*The Trauma of Birth*), in which, however, he claimed originality, it did not take long before his position was undermined. Groddeck and Rank were never expelled from the movement in the way Stekel and Adler were.

Briefly then, the way that we understand marginalization is as an interactional strategy used in strategic positioning. The decisive criterion to distinguish a marginal relationship from a more equal or nonmarginal relationship is, however, not the question of whether or not it ends in a formal break, but whether it follows a particular inner dialectic logic. We return to this logic of marginalization at the end of this chapter, after we have outlined some of the most important facts of Stekel's life and work as a psychoanalyst, and his troublesome relationship with Freud.

Wilhelm Stekel, the Unsung Populariser of Psychoanalysis

Little is known of Stekel's life other than the picture he himself outlined in his autobiography (Stekel, 1950a), which is both flattering and not entirely correct (Stanton, 1988). The editor of his posthumously published autobiography, Emil Gutheil, correctly judged: "The student of psychoanalysis can see in Stekel's notes how many of his own complexes remained obscure to him, can detect his unresolved narcissism, his overcompensated feelings of inadequacy; will smile when he reads that the man who was a master in ferreting out other people's repressions believed that he had hardly any himself" (quoted in Stekel, 1950a, p. 13).[2] But even so, the autobiography remains a useful, perceptive source for anyone who wishes to study the problem of marginalization.

Born March 18, 1868, in the town of Boyan in Bukovina (then part of the Austro-Hungarian Empire, later Romania), Stekel came from a middle-class Jewish family of merchants, with one elder brother and one elder sister. He went to the *k.k. Ober-Gymnasium* in Czernowitz, the capital of Bukovina, where, according to his own account, he developed a passion for reading and music, although apparently he was not a particularly good student. After the Gymnasium, Stekel went to Vienna to study medicine on a military scholarship; at the same time he also joined a pacifist movement. He believed himself to be a popular student, with a natural talent for learning languages and blessed as well with a musical and poetic gift. He worked for Krafft-Ebing for a while and settled as a general practitioner in Vienna. A first marriage produced two children: Erich-Paul Stekel (1898–1978), who became a musician, and Gertrud Stekel-Zuckerkandl (1895–1981), a painter. In the 1920s he separated from his first wife, Malvina Nelken, and became involved with Hilda Milko, whom he eventually married (Mühlleitner & Reichmayr, 1992).

During the latter half of the 1890s, Stekel published his first articles, both in medical journals such as the *Wiener Medizinische Wochenschrift* and in regular newspapers such as the *Neues Wiener Tagblatt* and the *Frankfurter Zeitung*, thereby earning himself a name as a journalist (Clark-Lowes, 1997; Stanton, 1988). Freud cited one of these medical articles (Stekel, 1895) in *The Aetiology of Hysteria* as evidence of the frequent exposure of children to "sexual assaults" (Freud, 1896, p. 207).

This citation created the first link between Stekel and Freud. The second came through Stekel's relationship with Max Kahane, an internist and early follower of Freud. Around the turn of the century, most likely in 1901, Kahane personally introduced Stekel to Freud.[3] Not long afterwards, Stekel turned to Freud for what he himself called a "sexual problem" (Stekel, 1950a, p. 107; see also Stekel's contribution on the history of the analytical movement in this volume). The exact nature of this sexual problem remains unclear, but it has been speculated that it was related to the fact that his first marriage was not a happy one (Stekel did not divorce his wife until 1938 but began to live with his future second wife shortly after the First World War) and that it involved sexual impotence, a pathological addiction to masturbation, or both (Roazen, 1975; Roudinesco & Plon, 1997).

[2] For some interesting biographical portraits of Stekel see Brome (1967), Freeman (1968), Ellenberger (1970), Roazen (1975), Rattner (1990), Mühlleitner & Reichmayr (1992), Jerábek (1994), and Castelnuovo-Tedesco (2002).

[3] Grosskurth (1991) dates the analysis in 1902, others, however, are inclined to date it much earlier, perhaps even as early as some time in the 1890s.

Indeed, in a letter from Freud to Wittels dated August 15, 1924, the masturbation hypothesis is hinted at almost explicitly when Freud wrote that one day "it will become clear that Stekel's claim about the harmlessness of masturbation is based on a lie," suggesting there was unpublished evidence indicating that Stekel was an onanist and that his errant behavior must be attributable to this practice. No such evidence has yet surfaced.

Stekel (1950a, p. 107) reported that his therapy lasted for "no more than eight sessions" (whereas Jones, 1953–1957, suggests that it lasted much longer). During this brief "therapy," Freud allegedly "expressed [...] surprise [...] that I had so few repressions" (Stekel, 1950a, p. 108)—a further indication, we believe, that Stekel's problems were indeed related to masturbation if this "absence of repression" is to be interpreted as being open about sexuality, for Stekel *was* strongly and most openly committed to the opinion, contra Freud, that masturbation was not harmful.

After his brief analysis with Freud, Stekel became an enthusiastic follower of his former therapist. He was in particular drawn to Freud's method of dream interpretation and liked the idea that everything can mean something different, and so he was quick to discover hidden symbolic meanings behind all sorts of symptoms. From this contact a relationship developed that was fairly intimate: Stekel was the only follower of Freud to address him by his surname instead of the formal "*Herr Professor*," used by everybody else (Jones, 1953–1957, Vol. II, p. 8). Freud recognized Stekel's intuitive gift for understanding dream symbolism and thought him serviceable to the cause, although he also believed that Stekel was theoretically weak. Furthermore, Stekel and Freud did not agree on many subjects. For example, Stekel emphasized psychic conflict (as in his theory of bipolarity) and he also opposed Freud's views on the origin of neurosis.

In 1902, it was Stekel who founded the weekly discussion group around Freud which would later become the well-known *Wiener Psychoanalytische Vereinigung* (Vienna Psychoanalytic Society) to which Alfred Adler initially belonged, as well as Paul Federn, Eduard Hitschmann, and Otto Rank, the society's secretary who began to keep notes of their meetings from 1906, and which were later published as "The Minutes of the Vienna Psychoanalytic Society" (Nunberg & Federn, 1962–1975; for brevity's sake referred to throughout this book as the *Minutes*). Over the years this small and informal round table discussion group evolved into a formal organization. In his 1926 historic account of the analytic movement, Stekel painted a picture of the first years of this society, in which, it appears, he himself played an important role.

Stekel was also one of the first to practice psychoanalytic therapy. After having abandoned his practice as a GP he set up a psychoanalytic practice in 1908. At this time he participated in all the events of the psychoanalytic society until he left it in 1912, like Alfred Adler before him, after many disagreements and difficulties with Freud.

While Stekel claimed that he had been rejected by Freud because the latter could not tolerate his disciple's scientific independence and creativity, Freud adhered to the conviction that it was because of his pupil's doubtful character. Be that as it may, a major subject of dissent lay in their respective views on masturbation: Freud thought that it caused neurasthenia, whereas Stekel believed it to be harmless and in fact argued that *repression* of the urge to masturbate caused neurosis. The problem of masturbation was a recurrent subject in the society's gatherings for an extended period of time (1907–1912) and twice led to a prolonged debate (in 1910 and 1912, respectively), one of which was published as volume II of the "Discussions of the Vienna Psychoanalytic Society" (Dattner et al., 1965). During these debates, Stekel's position proved to be entirely incompatible with Freud's. The debates effectively ended in Stekel's discontinuation of his involvement in the psychoanalytic society—or his being expelled from it. However, the question has been raised as to whether Stekel was expelled from the movement in 1912 because of his dissenting views, or whether it was the other way around: that his marginalization caused him to adopt a strongly antiFreudian point of view. We will return to this problem and discuss the motives behind this conflict in the chapter on masturbation and neurasthenia: Freud and Stekel in debate on the harmful effects of autoerotism.

It suffices for now to observe that from the moment Rank started to record the society's meetings, Stekel's position was already feeble. His presentations were often severely criticized and most of his publications received an even worse reception, as a result of which he sometimes felt "horrified at the lack of understanding which he encountered" (*Minutes*, Vol. I, p. 280). The Swiss in general, and Jung in particular, disliked him as a person, which was likely because they felt repulsed by his vanity. On the other hand, however, others judged him to be "a very agreeable companion," and, at least in comparison with Adler, "cheerful, light-hearted and very amusing" (Jones, 1953–1957, Vol. II, p. 135). Still, almost all of Freud's followers found it difficult to take him seriously. Stekel was considered to be a man "without the intellectual qualities to penetrate deeply" (Adler quoted in Andreas-Salomé, 1987, p. 35), or even to have "an irresponsible attitude toward truth" (Jones, 1959, p. 219). Few Freud biographies fail to mention that Stekel allegedly made up his own case histories (such as the infamous "Wednesday morning patient"—a patient he supposedly invented to prove that he had *just* seen a case like the one that was presented by someone else before the Wednesday evening society). Apparently, ridiculing stories such as these widely circulated in the Vienna society even before he broke with Freud, and Jones, who collected a good deal of them in his biography of Freud, used them to made it look as if Stekel had always been the laughing stock of the Freudians. But is it not possible to reverse the gaze and ask if humor could not have been part of Stekel's strategy of self-marginalization, and that this humor was used against him only *after* his break with Freud?

We return to this question in the last part of this chapter, and suffice for the time being to observe that Freud at least defended Stekel for a long time against attacks against his personality, although perhaps not always wholeheartedly. Thus Freud once wrote to Jung that Stekel's way of coming up with dream interpretations was similar to that of a swine who digs in the earth to find truffles. Similarly, he wrote to Binswanger that he could "still learn a good deal [from him]" (Freud to Binswanger, January 13, 1910). His book on dream symbolism in 1911 (*Die Sprache des Traumes*) was highly appreciated by Freud, who would, even after their break, write that it "contains the fullest collection of interpretations of symbols" (Freud, 1900, p. 375). Freud also hoped it would help popularize psychoanalysis. It did. Stekel gained considerable support outside Freudian circles—he was enormously popular in Austria, Germany, England, Holland, Hungary, Sweden, and the United States. Most of his books have been translated into several languages, and although his influence had by now declined, some of his works are still in print today. It was probably not without justification that he wrote that many students came to Freud through him (see the chapter: On the history of the analytical movement).

After his break with Freud, Stekel continued to call himself a psychoanalyst and persisted in his rebellious stance, even calling himself more Freudian than Freud. In order to introduce psychoanalysis to the medical profession he published ten large volumes that covered the whole spectrum of the "disorders of the instincts and the emotions" (*Störungen des Trieb-und Affektlebens*). Furthermore, he founded his own scientific journals and an institute for treatment, the *Institut für Aktive Psychoanalyse*, where, in contrast to the passive Freudian method, a form of "active psychoanalysis" was carried out. Not unlike Freud, he had his own circle of faithful disciples who worshipped his transcendent intuitive genius. Among his friends, supporters and students we find the following names: Owen Berkely Hill, A.S. Neill, Havelock Ellis, H.A.E. van Dishoeck (who translated some of his works into Dutch), J.L. Arndt, Samuel Lowy, Herbert Silberer (co-editor with Stekel of the short-lived journal *Eros and Psyche*), Fritz Wittels, S.A.Tannenbaum (to whom Stekel dedicated several works), Anton Missriegler, Fritz Wengraf, E. Bien, and Emil Gutheil (who were all collaborators on *Der Seelenarzt*, 1933; the latter was also, as noted above, editor of his posthumously published autobiography as well as several other works of his in English).

After the *Anschluss*, in 1938, Stekel, who was of Jewish origin, fled to Switzerland, then to England where he wrote his autobiography (Stekel, 1950a). An attempt to emigrate to the United States apparently fell through.[4] Like so many other refugees, he spent the next 2 years in London under difficult circumstances. When war broke out and a diabetic gangrene condition in the foot dete-

[4] In his *Rundbrief* (circular letter) of 25 June 1938 Fenichel reported how, according to Lawrence Kubie, some 57 colleagues applied for help to emigrate to America, "among whom, tragically, Stekel, who urges not to forget his merits for psychoanalysis." A.S. Neill also reported to Wilhelm Reich (May 6, 1938) that Stekel intended to emigrate to America (Placzek, 1981, p. 11).

riorated early in 1940, his living conditions became unbearable. He put an end to his life on June 27, 1940 (see Stanton, 1988 and Hilda Stekel's introduction to the autobiography of her husband, for an account of Stekel's last years in London).

Stekel wrestled with Freud's image until the very last days of his life. He admired the master, but at the same time it embittered him that Freud had abandoned him so abruptly and never allowed him to reconcile. He prided himself in calling himself a psychoanalyst, but believed at the same time that Freud had made "many mistakes." Upon receiving the news of Freud's death, Stekel entered a note in his autobiography that repeated in tone as well as in words what he had written 15 years earlier (in his 1926 account of the history of the analytical movement): "In my writing I have pictured the pettiness and the foibles of one of the greatest geniuses of our time, and shown how they influenced him in relation to myself; but the last thing I want is to produce the impression that I intended to belittle the greatness or deny the merits of this singular personality" (Stekel, 1950a, p. 285). But even if Stekel did not intend to upgrade his own image by downgrading that of the master, by criticizing Freud, he did link his own position to that of Freud. In a well-known metaphor that he began to use in the 1920s, Stekel said that since he was standing on the shoulders of a giant, he could see further than the giant—implying that his work was more Freudian than that of Freud. We return to this metaphor in our chapter on marginal historiography.

The disapproval and disregard of his works by orthodox psychoanalysts as well as their obliteration by the Nazis, did not favor the survival of Stekel's ideas in the German-speaking world. In the USA, however, where the rift between Freud and Stekel was never perceived as self-evident and where Stekel's active short-term analysis was bound to be favored above Freud's "passive," slow and therefore costly method, the popularity of his books has been a constant factor, only declining since the 1970s. Still, until recently neither in the German-speaking world nor in the United States did his ideas find much serious interest among historians. This "neglect" may be partly due to Ernest Jones' three-volume biography of Freud, in which, as we saw, many of Stekel's alleged shortcomings have been collected. And it is unwarranted because Stekel deserves recognition as one of the most fervent propagators of "the psychoanalytic gospel," not in the least through his articles in newspapers, of which many were reproduced in books. Moreover, his work demonstrates an interesting attempt at ameliorating (what he believed to be) shortcomings of the Freudian position with the help of his own insights and other approaches; he integrated useful observations and ideas of Pierre Janet, Carl Gustav Jung, and notably Alfred Adler (with whom he remained in contact after both broke away from Freud).

In the subsequent section we explore Stekel's position in psychoanalytic discourse in terms of the dimensions of marginalization discussed above, and try to answer in what sense it resembled that of optimal or constructive marginality.

Dimensions of Marginalization

There is great popular appeal to at least one aspect in the problem of marginalization, particularly with those who sympathize with the marginal: the idea that the marginal is deliberately pushed aside by someone with higher status. The marginal himself too often likes to complain about the unfair treatment his work has received. In point of fact, Stekel was of the opinion that his findings were completely ignored by the entire psychoanalytic community (*totschweigen* is the favorite word used in this context), although, he said, some of his most important ideas were smuggled in through the back door.

From the point of view of the marginal then, marginalization borders on the idea of conspiracy on the one hand, and on the problem of priority struggles on the other. Marginalization becomes a synonym for "repression" in an underdog discourse, used to designate the secret, illegitimate tactics of the oppressor, who not only puts the marginal down, but does not hesitate to steal from him either when he gets the chance.

Stekel (1926) himself mentions the idea of "bipolarity," which he insisted he had discovered independently of and before Bleuler (who called it "ambivalence"), and which was supposedly adopted by Freud without proper acknowledgement of his priority. Wittels (in Timms, 1995, p. 112) draws attention to death symbols in dreams, which he thinks Stekel recognized before Freud did (and introduced it as *Thanatos*). Even an unsuspicious source such as Andreas-Salomé (1987, p. 94) mentions how Stekel's concept of "polymorph criminality," a "colossal exaggeration" in her own opinion, was not without merit: Freud believed it points to anxiety about oneself.

So what are the indications that Stekel's work was actually "suppressed" from Freudian discourse, either before or after his break with Freud in 1912? And how does this neglect reflect on Stekel's position?

Prior to 1912, Stekel certainly was not repressed from psychoanalytic discourse, neither actively (in terms of his presence at the Vienna psychoanalytic society) nor passively (in terms of the number of citations of him, etc.). Up until 1912 Stekel was far from "neglected" or "ignored," but in fact a frequent speaker—second to Freud only—who presented more papers than any other member, and whose work was often cited and discussed by Freud, even if the references are often somewhat ambiguous in nature. Therefore, of the 141 meetings of the Vienna Society recorded by Rank for the period 1906–1912 (see Nunberg & Federn, 1962–1975), 16 were devoted to presentations by Stekel (as opposed to 14 by Sadger, 11 by Adler, 9 by Freud, and 8 by Hitschmann, with an average of only one to two presentations by other members).

After 1912, the amount of attention paid to Stekel's work diminished dramatically but it did not stop completely. His books were still reviewed and discussed in psychoanalytic journals and publications, albeit not in a positive sense. We mention a few sources. The *Internationale Zeitschrift für*

Psychoanalyse, the "official" organ of psychoanalysis, reviewed 10 of Stekel's works during the period 1914–1940; the *International Journal* 4. The popular *Psychoanalytische Volksbuch* devoted a section to Stekel in a chapter entitled "Dissenting courses and schools," in which his approach was typified as "pseudoanalytic," and his shortcuts to therapy (*Schnellverfahren*) criticized (Christoffel, 1939). *Entwicklungsziele der Psychoanalyse*, a small volume on recent developments in psychoanalysis by Ferenczi and Rank, published in 1922, made reference to only nine authors apart from Freud. One of them is Stekel, whose attribution of a specific neurotic symptom "first to sexuality, then to crime and finally to religion" was ridiculed in a footnote.

So it would be an illusion to believe that Stekel's work, his digression from or development of psychoanalysis went unnoticed; a fact that is corroborated by the recent publication of the circular letters (*Rundbriefe*) of Freud's secret committee, who anxiously watched Stekel's moves in the 1920s (Wittenberger & Tögel, 1999–2004).

This is to say then that the logic of marginalization does not follow a simple path of critical discourse which evolves to become hostile, or of a critique slowly being smothered or suppressed from official discourse (doxa). The main claim of this chapter is that a *dialogic* reading of the Freud–Stekel relationship attributes to a better appreciation of Stekel's marginalization, and, ultimately, of the various struggles in psychoanalytic historiography as well. But to appreciate the path of marginal dialogue, we need to study not only its interactional mechanisms or strategies, but also its progress in time. We outline two important interactional tendencies, and conclude with a general picture of the marginal route.

(a) *Originality-Priority*. While it is true that many of Freud's followers found it difficult to accept any improvement, adaptation, or alteration of psychoanalytic doctrine that did not come from Freud, Freud himself strongly induced his followers to explore new ways in a creative fashion. Therefore, he not only supported analytic applications to new territory and new problems, he also actively endorsed new theoretical and practical developments.

This, as Roazen (1986, p. 48) observed more than 30 years ago, gave rise to a particular problem, namely that "for a man really to be like Freud meant finally for him to be original. Yet originality ended his usefulness to Freud." This paradox conceals a particular dialogic principle which comes to the fore in a marginal's position: to insist on priority means to risk losing recognition for one's originality, but not to insist on it means risking losing recognition anyway. Many marginal relationships are therefore driven by a constant push and pull between these two factors. One way of solving this dilemma is to "go underground," by conducting the dialogue by implication: by referring to one's originality and priority only tacitly.

In our studies, we found that a large part of the Freud–Stekel dialogue was indeed implicit, and could only be understood by tracing back these embedded allusions. Indeed, many of the references to Stekel by other authors, including

Freud, do not mention him by name but only imply his work or position, and cannot be understood without a thorough knowledge of the context within which they appear. And conversely, many of Stekel's claims take as the point of departure Freud's implicit recognition for and critique of his work. Therefore, in the matter of masturbation, one of his major disagreements with Freud to which we devote a chapter in this volume, it is undoubtedly *because* Stekel launched his enlightened ideas in contradiction to those of Freud and his orthodox followers, who stressed the connection between masturbation and the development of neurasthenia, that he could have some influence. But a characteristic of this type of influence is that it could not be acknowledged as such. Also, in the chapter on Stekel's formative years (his *Lehrjahre*), we uncover some of the mutual influences between Stekel and Freud, in particular in the field of practice, by focusing on the element of autobiography and self-analysis. We identify a number of covert autobiographical passages in the work of Stekel, and attempt to link one of these "revelations" to an equally covert response to it by Freud.

So the first principle in a marginal dialogue is an implicit understanding, which is very much dependent on a local mutually shared context. This also reveals a weakness because such an implicit dialogue is obviously also very susceptible to distortions and misunderstanding. And to make explicit what is implied is also dangerous, for it disturbs the very foundations on which the dialogue is built.

(b) *Loyalty-Freedom*. Loyalty has always been an exceptionally important issue in psychoanalysis. To commit oneself to the cause was, for a while, sufficient reason for Freud to embrace the other as a "follower." In return, many of his students did their utmost to prove themselves "loyal" followers. But the question of loyalty runs persistently through not only Freud's works and those of his followers, but also through most of Stekel's, particularly those that deal with his own history, such as his autobiography and his 1926 account of the history of the analytical movement, which reads as one long attempt to be acknowledged as a loyal yet independent student. This tension may lead to a particular chain of interaction that can easily become a negative spiral, when the marginal attempts to enhance more recognition for his work by increasing his dissentient viewpoints, but as a reward is marginalized even more.

This is true in the case of Stekel, who claimed that he had developed an independent position but still insisted on being called a psychoanalyst. He prided himself in having opposed Freud on a number of important issues, such as the question of whether or not masturbation causes neurasthenia, but at the same time believed that he was more Freudian than Freud. He also claimed to have formulated several ideas that were only much later, and without proper acknowledgement, introduced into psychoanalytic theory, such as the idea of a death-wish.

Stekel, it should be stressed, had a strong popular appeal. His work was influential in lay as well as in medical circles and was recognized as "original," perhaps even as "refreshing," and it was certainly judged to be different from that of Freud. It is also said to have helped transform society's predominant distaste for the medicalization of deviant sexual practices, including homosexuality. Thus Caprio (1954, p. 112) wrote that Stekel was "an original co-worker of Freud's, [who] made an important contribution to the understanding of homosexuality by further developing the theory of bisexuality" (see also London & Caprio, 1950; Wolff, 1966).

Therefore, the question in a marginal relationship is for the marginal to know how to attain a degree of independence from and yet sufficient commitment to the dominant, and for the dominant to know how to keep enough distance from the marginal without having to renounce him (a problem that was in many ways already pointed at a long time ago by Norbert Elias (1956)). In some respects the marginal and the dominant seem condemned to continue their struggle forever, although in other respects they may develop a degree of (temporary) equilibrium. However, the constant wavering between the two extremes—loyalty/freedom—in such a symbiotic relationship, calls for a resolution that can easily take the form of appealing to the audience for sympathy or understanding.

This characterizes the second element in the marginal dialogue, which is diametrically opposed to the first, namely the tendency to expose or reveal the dynamics of the marginal relationship, and the hidden bonds that bind them. Both Stekel and Freud at times tried to solve the problem of loyalty and freedom in their antagonistic relationship by publicly revealing particularities about it. But the effect of these "confessions," as could be expected, only reinforced the tension.

With these two opposing tendencies briefly described, let us now turn to the final characteristic of the marginal relationship: its phased progression in time, or current development. For the Stekel–Freud relationship, we find that there are five phases to be recognized, each characterized by its own particular dialogical or interactional pattern.

(a) *Initial Stage.* Stekel comes to Freud as a patient and goes through brief analysis. In our chapter on Freud and Stekel in debate on the harmful effects of masturbation, we trace the first form of marginalization in their relationship to the moment when Stekel attempts to counter-analyze Freud and Freud refuses to be analyzed. The dialogical characteristics of this relationship (as far as it can be reconstructed, since almsot no letters from this period have survived, and no implicit or explicit references to either Stekel or Freud have been identified) basically take the form of a training or instruction, based on asymmetrical distribution of power/ knowledge.

(b) *Second Stage.* Stekel becomes an adherent of Freud, writes pamphlets to propagate psychoanalysis, reads Freud's works, and initiates the Wednesday evening society; Freud writes a preface to his anxiety book. This period, which lasts until approximately 1908, is characterized by what could be called "positive marginalization." Stekel stands out as a rebellious but original follower, whose works are occasionally referred to. Their relationship at this point in time may be described as following a dialectic of antagonism, as we have pointed out in the chapter on the making and breaking of Stekel as an analyst. By and large, this means that much of the differences between Stekel and Freud do not surface but remain underground, but also that their debates contribute to the construction of a positive identity in the works of each other, and to the development of psychoanalytic knowledge in a broader sense.

(c) *Third Stage.* Stekel publicly opposes Freud; Freud ceases to refer to Stekel. This period, which lasts until approximately 1910, marks a reorientation of both Freud and Stekel toward each other, in particular with respect to the public. The positive marginalization gives way to a negative type in which the marginal's position is no longer perceived (by either party) to be constructive. The dialogical characteristics may perhaps be best described as taking the form of denial, either by open negation or publicly ignoring the other.

(d) *Fourth Stage.* Stekel is becoming a *persona non grata*; his rebellious position is no longer recognized as legitimate, his expulsion from the movement becomes inevitable as he opposes Freud in ever stronger terms. This period ends in 1912, when Stekel ceases to be a marginal in psychoanalysis and becomes an "independent." The type of marginality that emerges during this time could be called destructive marginality, as it clearly moves toward the breaking off of relations. There is little communication during this period.

(e) *Fifth Stage.* All relations between Freud and Stekel are broken off; there is no longer any communication between them whatsoever. But interestingly, both still wrestle with each other's images, and consequently the dialogic characteristics are now open hostility. The type of marginalization that belongs to this stage is the oppressive type.

Conclusion

To call Stekel a marginal is self-evident for anyone who knows even only a little about him, yet it is also problematic. In what sense was his role different from that of Jung? Or of Groddeck? Or of Adler? Indeed, of *any* psychoanalyst at that time? Were they not all marginals in one way or another? After all, was not psychoanalysis a marginal practice itself, that never succeeded in finding acceptance anywhere in the academic world?

These questions, legitimate though they are, cannot be answered within the scope of this book. We have confined our approach to one person, one practice, one period in time. But even within the confinements of this approach we were stunned by the richness of the case, and we hope that it stimulates others to reapproach the marginals' history—not because we think they need to be rediscovered and assigned to a new place, but because we are of the opinion that an understanding of their position helps us to understand the position of others.

Stekel's 1911 book *Äskulap als Harlekin,* published under the pseudonym of "Dr. Serenus," contains a disguised critique of Freud.

SUSQUEHANNA 7-7510
ENDICOTT 2-6100

MRS. HILDA STEKEL
40 WEST 72ND STREET
NEW YORK 23. N. Y.

June 30,1960.

Mrs.Helen Walker Puner
c/o Dell Publishing Co., Inc,
750 Third Avenue
New York.N.Y.

Dear Mrs. Walker Puner,

I read with greatest interest your biography of of Freud. I agree with Erich Fromm,who writes in his foreword:-"Her book is more analytical and realistic than those written by psychologists so far."

You were unbiased what you wrote about my late husband,Dr.Wilhelm Stekel.However I feel and so do many others,who greatly enjoyed your book,that it is a pity that you wrote only about his beginnings as a psychoanalysts.His real development happened later. But since there is the tendency in certain psychoanalytical circles not to mention Stekel,it is rather difficult to be informed about him .You may therefore not know about his later work. I should like to tell you briefly about it.

He developed his own technique and thus he became the founder of the active method of analytical psychotherapy. As you know brief analytical therapy is much practised nowadays.

Amongst his important publications is his work Disorders of the Instincts and the Emotions 10 vls. It was published here as well as his last book on dream interpretations and others of his works.

Since your book had such a tremendous success,I assume,it will soon have a second edition. I would very much appreciate if you considered to write about Stekel's development. You would find valuable information in his Autobiography,Publishing Corporation Liveright 1950., Technique of Analytical Psychotherapy ,revised edition by Bodley Head,London. I think,it would be of interest to you,to read (if yo can read German) his paper on'Die Geschichte der Analytischen Bewegung' published in Die Fortschritte der Sexualwissenschaft and Psychanalyse 2.Bd Herausgeber Wilhelm Stekel,Leipzig und Wien,Franz Deuticke 1926. I am sure one can get it in the National Library.

If you should take up my suggestions,I will be gla to help you with information.Perhaps,I should mention that I worked with Stekel as his assistent and collaborator for 23 years. This matter is the fore very close to my heart.

Realizing how very busy you may be, I hope you won't mind,that I took up so much of your time.

May I thank you very much for the pleasure I derived from reading your stimulating book.

Yours sincerely,

Hilda Stekel

Letter of Hilda Stekel to Helen Walker Puner, author of *Sigmund Freud, his Life and Mind,* dated 30.6.1960, on the importance of, among other things, her late husband's autobiographical study "On the History of the Analytical Movement". Letter in possession of the authors.

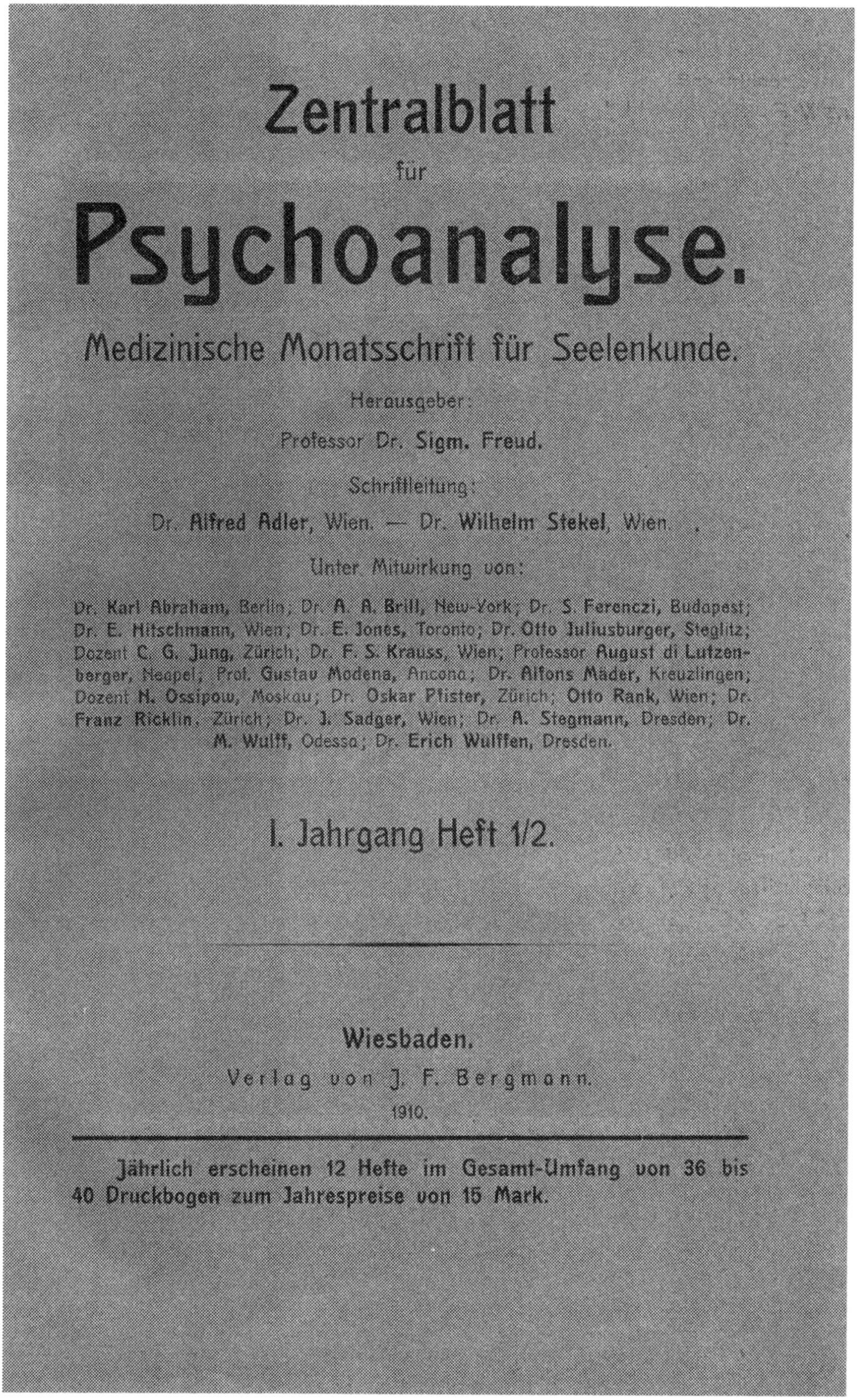

Zentralblatt

für

Psychoanalyse.

Medizinische Monatsschrift für Seelenkunde.

Herausgeber:

Professor Dr. Sigm. Freud.

Schriftleitung:

Dr. Alfred Adler, Wien. — Dr. Wilhelm Stekel, Wien.

Unter Mitwirkung von:

Dr. Karl Abraham, Berlin; Dr. A. A. Brill, New-York; Dr. S. Ferenczi, Budapest; Dr. E. Hitschmann, Wien; Dr. E. Jones, Toronto; Dr. Otto Juliusburger, Steglitz; Dozent C. G. Jung, Zürich; Dr. F. S. Krauss, Wien; Professor August di Lutzenberger, Neapel; Prof. Gustav Modena, Ancona; Dr. Alfons Mäder, Kreuzlingen; Dozent N. Ossipow, Moskau; Dr. Oskar Pfister, Zürich; Otto Rank, Wien; Dr. Franz Ricklin, Zürich; Dr. J. Sadger, Wien; Dr. A. Stegmann, Dresden; Dr. M. Wulff, Odessa; Dr. Erich Wulffen, Dresden.

I. Jahrgang Heft 1/2.

Wiesbaden.

Verlag von J. F. Bergmann.

1910.

Jährlich erscheinen 12 Hefte im Gesamt-Umfang von 36 bis 40 Druckbogen zum Jahrespreise von 15 Mark.

First issue of the *Zentralblatt*, edited by Stekel and Adler under the directorship of Freud, which would come to play an important role in the dissension between Freud and Stekel.

NERVÖSE

ANGSTZUSTÄNDE

UND IHRE BEHANDLUNG

VON

DR. WILHELM STEKEL

WIEN

DRITTE, VERMEHRTE UND VERBESSERTE AUFLAGE

Hinter jedem Gedanken lauert eine Leidenschaft . . .

August Strindberg

URBAN & SCHWARZENBERG

BERLIN N
FRIEDRICHSTRASSE 105b

WIEN I
MAXIMILIANSTRASSE 4

1921

Third edition of *Nervöse Angstzustände* (Nervous Anxiety States), Stekel's most well-known book for which Freud would write a preface, appears here as the first volume in the "Disorders of the Instincts and the Emotions"-Series.

Wilhelm Stekel at the Weimar Congress, 1911. Detail of a group portrait, courtesy of the Freud Museum, London.

Chapter 2

A SILENT ANTIPODE
The Making and Breaking of Psychoanalyst Wilhelm Stekel[1]

Jaap Bos

Friend by enemy I call you out.
Dylan Thomas

Introduction

In the previous chapter we mentioned briefly that Stekel met Freud around the turn of the century, underwent brief therapy and not long afterward became an enthusiastic follower, indeed, a zealous popularizer of psychoanalysis. But we also noted how history records Wilhelm Stekel as a somewhat tragic figure: one of Freud's very first followers who was one of the first to fall from the master's grace after having behaved in a manner "of which it is not easy to publish an account" (Freud, 1914, p. 47) and was never permitted reentry into the community. Descriptions of Stekel typically mention two "character flaws," of which his alleged lack of scientific integrity is the gravest. According to Jones (1953–1957, Vol. II, pp. 134–135), "he had no critical powers" and "no scientific conscience" at all. Stekel is said to have made up case histories in order to illustrate or corroborate his own ideas—talk of his "Wednesday-evening patients" is proverbial in this respect. The other flaw concerned his disagreeable nature. He allegedly had an inflated self-image (Brome, 1967, p. 66), showed exhibitionistic pleasure in disclosing private affairs (Roudinesco & Plon, 1997, p. 1010), and took a quasipornographic interest in sexuality (Roazen, 1975, p. 211); in short, he seems to have

[1] This chapter first appeared in *History of Psychology*, 6(2003), 331–361.

been an "unbearable" human being whom Freud eventually came to judge as "morally insane" (Freud to Ferenczi, cited in Jones, 1953–1957, Vol. II, p. 137). Wittels wrote that "Freud could not bear the sight of him [Stekel] and did not miss a single opportunity to point this out verbally or in writing" (in Timms, 1995, p. 118). Even more positive descriptions of his personality are usually accompanied with condescending coda: "a fluent if careless writer, a born journalist in a pejorative sense" (Jones, 1953–1957, Vol. II, p. 135); "a talented and prolific writer, as well as a poet and an excellent musician [...] but his work was somewhat journalistic" (Roazen, 1975, p. 211).

Stekel was a productive writer indeed. In the 28 years between his break with Freud and his death in 1940, he remained active as a psychoanalytic psychotherapist and author, publishing at least one book a year as well as numerous articles. Grinstein (1958) listed an astonishing 367 titles to Stekel's name—36 books or booklets, 179 articles, and 153 abstracts and reviews—and this list is far from complete. In contrast to Freud, however, Stekel was first and foremost a clinician; *Störungen des Trieb-und Affektlebens* (Disorders of the Instincts and the Emotions), published in 10 volumes between 1912 and 1928, celebrates the psychoanalytic case history and is considered to be his magnum opus. It was remarkably successful. Most volumes have been translated into English and French and have been reprinted at least twice in these languages, some many times more. The most recent reprint dates from 2004 (a reprint of his book on masturbation (Stekel, 1950c)).

If Stekel's pre-1912 writings are considered "valuable" (Jones, 1953–1957, Vol. II, p. 134), the majority of his post-1912 writings are, however, almost entirely neglected by Freudian psychoanalysts and historians. Many believed that after the break with Freud Stekel's work deteriorated into little more than speculation as his intuitive understanding of psychoanalysis no longer stood under the stern influence of reason.

All of this is to say that for many years the predominant picture of Stekel was that of an almost pitiable character, whose suicide in 1940 only added to his tragedy.[2] Recently, attempts have been made to adjust aspects of this image (Clark-Lowes, 2001; Kuhn, 1998; Nitzschke, 1992), particularly in the field of pedagogy, where his influence was large (see Groenendijk, 1996; 2001), but in mainstream psychoanalytic historiography, Stekel still appears as someone with whom Freud put up with for so long because in the beginning there were no "decent followers"—someone whose influence on the development of psychoanalysis was minor and whose work is virtually without merit.

[2] After the occupation of Austria, Stekel balked at leaving Vienna and only just managed to escape the Nazis in 1938. He ended up in London, where he ended his own life in 1940. His self-chosen death has been put on a par with the suicides of two of his friends, Kahane and Silberer (see Roazen, 1976; Sulloway, 1979), but Stekel ended his life after prolonged suffering from a severe disease. His last years more closely resemble those of Freud, who also only just managed to flee Austria after its occupation and also ended his life by his own hand after prolonged and dreadful suffering (see Stanton, 1988).

We believe this image to be unfair in many respects. True, Stekel's work does not compare to Freud's, but neither does that of any other of his followers; and with regard to his character flaws, it is more likely that Stekel was utterly vain but not morally insane. Nonetheless, we do not seek to reevaluate his contribution, for we have neither the skill nor the desire to undertake such a project. Nor is it our aim to recast Stekel's position in the history of psychoanalysis.[3] The aim of this chapter is limited to a reassessment of Stekel's rise and fall in the psychoanalytic community. By examining this process in detail, we hope to make clearer how and by what discursive mechanisms the early socialization of psychoanalysts took place. Contrary to what is sometimes believed, what held Freud's earliest followers together was much less a set of shared philosophical or psychological assumptions, let alone a specific generic technique, but much rather what may be called a particular *discursive attitude*, a certain commitment, perhaps, or moral obligation toward Freud and "the cause" in general. By exploring the process of the making and unmaking of Stekel as a member of the psychoanalytic community, we intend to make visible the discursive dynamics of this general attitude.

The structure of this chapter is as follows: we begin with a general outline of the problematic student–teacher relationship between Freud and Stekel, focusing on the latter's role as silent antipode, and then closely examine several episodes in their working relationship. In particular, we focus on a hitherto underappreciated aspect of Freud's *Gradiva* study (Freud, 1907a) in relation to one of Stekel's major early works, *Die Träume der Dichter* [The Dreams of Poets], namely, its almost explicit instructional character. Next, we discuss Stekel's contributions to the debates and discussions in the Vienna Psychoanalytic Society. At the chapter's conclusion, we return to the question asked at the beginning: What are the constituent factors in the making and unmaking of Stekel as a psychoanalyst?

"I Was His Christ, He Was My Judas"

The starting point of Stekel's relation to Freud is very likely his therapy which lasted "no more than eight sessions" (Stekel, 1950a, p. 107). In his 1926 account of the history of the analytic movement he calls his reasons for going into analysis "confusions" (*Wirrnisse*), and it is quite likely that these confusions had to do with masturbation. Elsewhere he wrote that as a child he read a book by Dr. Retau, warning against this practice, and that this caution caused him a great deal of problems, which he only overcame when he no

[3] However, we believe it is time his position be reconsidered. Evidence, some of which is presented in this and subsequent chapters, indicates that Stekel's role in the early history of psychoanalysis was more important than is commonly known.

longer repressed the urge to masturbate. During this brief therapy, Freud allegedly "expressed ... surprise ... that I had so few repressions" (Stekel, 1950a, p. 108), an indication, we believe, that Stekel's problems were indeed related to masturbation if this "absence of repression" is to be interpreted as being open about sexuality. In later years, Stekel strongly and most openly committed himself to the opinion, against Freud, that masturbation is not harmful. We return to this matter later.

What Freud and Stekel discussed during Stekel's brief therapy also remains the subject of speculation, but it may well have been focused on a mother theme, because Stekel mentioned that at one time Freud interpreted one of his dreams as incestuous (a "mother fixation"), and Walter Schindler, one of Stekel's pupils, claimed that Stekel's mother was "a strong personality who played a leading role in his life" (Schindler, 1980, p. 7).[4]

Although Stekel appears to have been enthusiastic about Freud from the moment he met him, he did not accept this interpretation and shortly afterward discontinued therapy (Freud to Jung, 22 Dec. 1912). He was nonetheless one of the first to wholeheartedly support Freud's *The Interpretation of Dreams* in public, praising its author in a long review as a "genius," an "expert on sick souls" [*den berühmten Kenner der kranken Seele*], whose method of dream interpretation provided an "important therapeutic means." However, Stekel's reading of *The Interpretation of Dreams* was already biased by his own expectations in the sense that he believed that "Freud's method of dream interpretation is very simple" [*sehr einfach*] (Stekel, 1902/1986, p. 113) and led to instant understanding. He was thus no slavish follower and in fact criticized Freud for his heavy emphasis on the repression of sexuality.

Neither this doubt nor his biased expectations prevented him from becoming a disciple of Freud, and soon he began writing many more articles promoting Freud's ideas and, even more important, initiated the meetings of the "Psychological Wednesday Society" (the forerunner of the Vienna Psychoanalytic Society) in 1902. "I was his apostle, he was my Christ," Stekel (1950a, p. 106) wrote, in his own somewhat pathetic style. He would therefore become one of the first of Freud's supporters, and he was certainly one of the first, if not *the* first, to practice psychoanalytic therapy after Freud (there are more claims to that title, and clearly it is difficult to mark psychoanalytic therapy's exact beginning, even for Freud).

Stekel's relations with the master, as we have mentioned, were antagonistic from the start and colored by many personal as well as scientific differences. In fact, all of his pre-1912 work explicitly digresses from Freud

[4] Jones (1953–1957, Vol. II, p. 7) had the impression from Freud that Stekel's therapy lasted "much longer" than a mere 8 or 10 sessions, and he suggested that he was aware of the precise nature of Stekel's problem.

to some degree. *Die Ursachen der Nervosität* [The Causes of Nervousness] (Stekel, 1907a), for example, was published in 1907 and bears a dedication to Freud. In this booklet, Stekel claims to be proud to call himself "a pupil of Freud" but, he wrote, "that should not mean to say that everything which I say corresponds with [Freud's] opinions. On the contrary! I'm quite sure that in many places he will shake his head, as if giving vent to his dissatisfaction." When *The Causes of Nervousness* was discussed at the Vienna Psychoanalytic Society on December 5, 1906, it was indeed dismissed by most discussants, including Freud, who believed Stekel remained arrested at the point of his own *Studies on Hysteria* (Freud & Breuer, 1895b). One opponent, Rudolf Reitler, went as far as to describe the book as a "neurotic symptom"; Stekel admitted to "having written the booklet in a state of severe psychic depression" (Nunberg & Federn, 1962–1975, Vol. I, pp. 71–73).

Stekel's next publication, *Nervöse Angstzustände und ihre Behandlung* [Nervous Anxiety States and their Treatment] played an important role in his formation as an analyst. This piece first appeared as an article in the *Medizinische Klinik* (Stekel, 1907b), then as a separate book (Stekel, 1908), and finally as the first volume of his *Disorders of the Instincts and the Emotions* series (Stekel, 1912–1928). Chapters were presented to the Vienna Society as separate case studies on hysteria, psychosis, and impotence.[5] Stekel (1950a, p. 120) considered it "perhaps my best book," and he supplemented each subsequent edition with fresh material so that eventually a ten-page article grew into a 600^{+}-page hodgepodge of his own and others' case histories by its third edition in 1921. Although Stekel's repeated enhancements to the work followed the example Freud set in *The Psychology of Everyday Life* and *The Interpretation of Dreams*—both of which had also been substantially expanded with observations by others (including Stekel) in each new edition—he had in fact gone even further inasmuch as he had also closely copied some of Freud's arguments in detail. In one instance, in which he narrated how he once forgot (repressed) the names of two Italian (!) patients and explained the associative paths that eventually led him to understand the reasons for forgetting them, he literally rephrased almost word for word Freud's famous Signorelli argument, in form as well as content, even including the figure that graphically represents the entire argument (Stekel, 1921a, pp. 644–645).

Although Stekel thus followed Freud in some ways, he also explicitly opposed him in other, important respects. From the very start, he had considerable difficulty in accepting Freud's differentiation between actual neurosis (caused by dissatisfying sexual conditions in the present) and psychoneuroses (rooted in the psychosexual development), and even

[5] Compare Nunberg & Federn (1962–1975), Vol. I, minutes 30, 31, 35 and 50, and Stekel (1909), chapters 22, 26, 34, and 25.

more so between *anxiety neurosis* (belonging to the group of actual neurosis) and *anxiety hysteria* (belonging to the group of psychoneuroses). Nevertheless, he himself introduced the latter term in his work on nervous anxiety states (Stekel, 1908). Stekel later revealed, however, that this was at Freud's explicit instigation; Freud allegedly told Stekel: "I will give you a royal present; we shall call all cases where the anxiety has a psychological root 'anxiety hysteria,' while cases where anxiety can be traced to injuries of the sex life will be called 'anxiety neurosis'" (quoted in Stekel, 1950a, p. 137).

Stekel accepted the "present" but used the term *anxiety hysteria* to describe nervous conditions that are rooted in what he called a psychological *conflict*. The conflict, he said, is caused by a tension between two opposing drives: the life instinct and death instinct (*Minutes*, Vol. I, p. 175). From this struggle or conflict Stekel would come to what he considered a "psychic law": the law of bipolarity, that is, an eternal struggle between two opposing tendencies. He thus already tended to reduce the etiology of all neuroses to psychological conflicts.

As with his previous work, Stekel's prepublication presentations of *Nervous Anxiety States* failed to find a warm welcome in the Vienna Psychoanalytic Society, even though Freud himself was much less hostile toward it. Freud tolerated Stekel's use of *psychological conflict* and *bipolarity*, but he believed these notions to be a mere rephrasing of his own concept of repression (*Minutes*, Vol. I, p. 72). He also objected to Stekel's simplification of his theory of psychosexual development (hence his remark that Stekel practiced "surface psychology"; *Minutes*, Vol. I, p. 395) and above all to the claim that *all* neuroses are caused by a psychological conflict. While not taking an altogether negative stance, much of Freud's attention in the Vienna society with regard to Stekel was devoted to attempts to make him more aware of the diagnostic differences between hysteria, neurosis, psychosis, and so on. However, Stekel would never come to accept the differentiation between actual and psychoneuroses, and he dropped the entire nomenclature, including the "royal present," immediately after his break with Freud, claiming that "the differentiation between anxiety neurosis and actual neurosis contradicts the experiences in psychotherapeutic praxis. [...] There are no actual neuroses in the Freudian sense. There are only psychoneuroses and in fact just one type of psychoneurosis with various outer forms and in varying degrees of intensity" (Stekel, 1909b, p. 25).

Stekel claimed the book edition of *Nervous Anxiety States* was "an immediate success" (1950a, p. 120). Freud, who contributed a foreword to this manifestation of Stekel's work, did not think ill of it, although, as we mentioned, he had his doubts. He recommended it to Abraham (Freud to Abraham, 16 Feb. 1908) and told Jung that the book was "theoretically weak, but rich in skilful analyses and likely to make an impression" (Freud to Jung,

14 Jan. 1908). Jung, however, was not at all impressed. He valued the case material but regretted the "incompleteness of [Stekel's] analyses" and believed many of the interpretations to be "very arbitrary" (Jung to Freud, 11 Aug. 1908). Freud disagreed. True, Freud wrote, Stekel had an "uncritical mind" and an "incorrigible tendency to content himself with approximations," and he was "sloppy," but "all the essentials are correct." "This author's gift for psychoanalysis allows him to make with impunity mistakes that would be more harmful in the work of another, less gifted man" (Freud to Jung, 13 Aug. 1908).

Freud thus took an ambivalent stand toward Stekel. He no doubt harboured an "initial sympathy" for him, as he later admitted. "Perhaps you still remember how our relationship began?" he asked, years after their relationship had turned sour (Freud to Stekel, 13 Jan. 1924). Although Freud never held a high opinion of Stekel as a scientist and despised his "unbearable manners" (Freud to Wittels, 18 Dec. 1924), it did not prevent him from trying to turn Stekel into a useful tool for the cause. Stekel's 1908 book *Nervous Anxiety States* could not satisfy him personally, he said, but he thought it would "do immense good among the outsiders, his level being so very much near to theirs" (Freud to Jones, 20 Nov. 1908). Freud also commissioned Stekel in 1910 to form a committee whose task was to collect symbols and typical dreams (Freud to Ferenczi, 25 Feb. 1910),[6] some of the results of which were published in *Die Sprache des Traumes* [The Language of Dreams] (Stekel, 1911). When Jung heard of this plan, he exclaimed: "Good lord! That's all we needed!" (Jung to Freud, 15 Nov. 1909), but Freud replied: "A book on dream symbolism doesn't strike me as impossible," although he added, "but I'm sure we shall object to the way Stekel goes about it" (Freud to Jung, 21 Nov. 1909). What he did not write to Jung was that the plan had come from Jones, and that he asked Stekel to carry it out.

Freud protected Stekel, through thick and thin, for more than 10 years. When, for example, Ludwig Binswanger complained to Freud about Stekel, he was told that he could "still learn a good deal [from him]" (Freud to Binswanger, 13 Jan. 1910), to which Binswanger humbly admitted: "One could indeed learn much from Stekel, once one had passed a certain point" (Binswanger to Freud, 9 April 1910).

Stekel, for his part, was as loyal to Freud as he could possibly be without ever giving up his own ideas or his critique. When he became editor of the *Zentralblatt* in 1910, he contributed numerous reviews and psychoanalytic observations (*Mitteilungen*), all of which were purported to support Freud's general theoretical framework, yet at the same time he also promoted his own pet ideas, and Freud soon began to feel uneasy about the disproportionally large amount of "Stekeliana" in the journal. It certainly seemed as if Stekel considered the *Zentralblatt* more or less as *his* periodical, as Freud

[6] The original proposal came from Jones (see Jones, 1953–1957, Vol. II, p. 68).

complained. It is perhaps more appropriate to say that (at least initially) Freud had let him believe this to be the case because, as legend has it, Stekel and Adler were appointed editors of the journal after a dramatic confrontation in Stekel's hotel room at the 1910 Nuremberg congress as a "compensation" for their "lost hegemony" when Jung, rather than a Viennese, was elected president of the International Psychoanalytic Society (Freud, 1914, p. 45).

But Stekel's 1911 *The Language of Dreams* upset Jung considerably. Although not a dictionary, the book categorized the most "typical" dreams and dream symbols from a Freudian perspective. Again, however, it was by no means a strictly Freudian book, because the case material presented in it is treated in a more Stekelian than Freudian manner with the interpretations being extremely clear cut if not obvious and invariably related to a psychic conflict. In addition, his interpretations always pointed toward the immediate resolution of the conflict. From the very beginning, Stekel was an outspoken advocate of short-term treatment.

Although characterized by his colleagues as cursed with cheap, shallow interpretation, theoretically weak and full of ideas that were not nearly as elaborate as Freud's, Stekel was nonetheless a commercial success. By 1911, he had already written a number of popular books and newspaper articles on various aspects of applied psychoanalysis for the masses, and he had earned himself a reputation such that he was taken by some as a representative of Freud's followers in general. Karl Kraus, for example, made "Stekel, not Freud, ... the central target of that great assault on psychoanalysis [...] In Stekel he attacks the journalistic vulgarisation of theories which he had once admired" (Timms, 1986, p. 110). Stekel's reputation displeased Freud and the rest of the early psychoanalysts, such as Bleuler and above all Jung and Abraham (who also despised Stekel).

It is not surprising that tensions between Stekel and Freud grew over the years but they were already clearly visible prior to the *Zentralblatt* affair, which finally sealed Stekel's fate. Indeed, trouble between the two had peaked two years earlier, in the first months of 1911, when Adler was forced to leave the movement and Stekel threatened to leave with him. Freud struck back with a review of Stekel's *The Language of Dreams* (Stekel, 1911c) that was apparently so vicious that Bleuler refused to print it in the *Jahrbuch*. (Freud promised to rewrite it, but the review has never surfaced.)[7] Then, in the summer of 1911, came a reconciliation that lasted until November, at which point Herbert Silberer submitted an article that Freud believed was "cooked up" by Stekel: "You can't imagine how he poisons the whole undertaking for me.

[7] However, a summary of it is found in *The Minutes of the Vienna Psychoanalytic Society* (Nunberg & Federn, 1962–1975, Vol. III, p. 236). From it we gather that Freud's main objections were against Stekel's excessive reliance on symbolic interpretations and against his overvaluation of the manifest dream content.

Every time I think of him the old litany come to mind: 'Me piget, Puget, poenitet, taedet, taque miseret' [I feel disgust, shame, regret, weariness and pity]. But I have decided to get along with him" (Freud to Jung, 12 Dec. 1911).

The final blow came a year later, in October 1912, when Stekel refused to let Tausk take over the review section in "his" *Zentralblatt*. Freud demanded "unconditional obeisance" [*volle Unterwerfung*] (Freud to Ferenczi, 20 Oct. 1912), and when Stekel refused to surrender, Freud dropped both Stekel and the periodical. Although Stekel made it appear that he had convinced the publisher, Bergmann, to side with-him against Freud, the reverse was true: Bergmann charged Freud with breach of contract and forced him to pay 650 DM.

In November 1912, Stekel officially withdrew from the Vienna society. A month later Jung, as the president of the *Internationale Psychoanalytische Vereinigung*, wrote to the local presidents of the society to inform them of the loss of the *Zentralblatt* and the simultaneous founding of a new periodical, the *Internationale Zeitschrift*. Meanwhile, Freud asked his followers to withdraw their support for the *Zentralblatt*, which most did. When the next issue appeared, only Stekel's name appeared on the cover, rather than the customary list of some 30 names (although a few, such as Schilder and Silberer, continued to write for the journal). The journal was published for another year and a half and was discontinued at the outbreak of the First World War. The events that led to this break are described more fully in chapter 7.

Now that Freud was free of Stekel, some ambiguity, but clearly not all, left Freud's tone. On the one hand, he claimed to be "extremely happy" [*Gottesfroh*] to have gotten rid of him (Freud to Binswanger, 28 Dec. 1912) and now believed that this loss was in fact a gain (Freud to Putnam, 1 Jan. 1913 in *Letters of Sigmund Freud*, p. 299). From that point, everyone within the psychoanalytic community knew that anyone even associated with Stekel was suspect. Thus, for example, when Andreas-Salomé intended to publish an article in Bloch's *Zeitschrift für Sexualwissenschaft*, she first inquired whether Freud would object, as "St. Ekel" frequently published in that journal (*Ekel* is German for "disgust"; Andreas-Salomé to Freud, 18 Dec. 1916). In letters and conversations Freud invariably told his audience how he had once stood by Stekel "in the face of attacks from all sides" and how "treachery and ugly dishonesty" forced him to push Stekel aside (Freud to Wittels, 18 Dec. 1923, in *Letters of Sigmund Freud*, p. 346). At one point Freud is even reported to have said that he committed two crimes in his life: "I drew attention to cocaine and I introduced Stekel to psychoanalysis" (Wittels, quoted in Timms, 1995, p. 123).[8] If Freud had ever seen Stekel as his apostle, it must certainly have looked now as if he were his Judas, and although he could not bring himself to read Stekel's publications he would ask his followers to do

[8] It is perhaps not a coincidence that in this comparison Stekel is put on a par with an addictive drug, because to Freud all addictions go back to infantile masturbation.

so for him. Freud and his followers continued to watch Stekel's every move; Stekel was gone but not forgotten.[9] Even after their break, Freud did not cease to refer to Stekel's work in a remarkably positive and affirmative sense, as we show later.

Stekel never ceased to admire Freud either, and for a long time harboured the hope of being allowed to return to the master, believing his own contribution to psychoanalysis invaluable.[10] His statement that he who is standing on the shoulders of a giant can see further is well known; Freud's ironic reply, that a louse on the head of an astronomer cannot, is no less frequently quoted.[11] Stekel is reported to have told Joseph Wortis (1954, p. 142), who was in training analysis with Freud at that time, that "Freud is one of the greatest of geniuses," but Freud "rejected the compliment by saying it was purposely meant to reach his ears." Stekel's admiration for Freud eventually gave way; in *Technique of Analytical Psychotherapy* (Stekel, 1950b, p. xx), he wrote in an embittered tone about his former mentor and confided that "after thirty years" experience of analysis, "I no longer believe in the overwhelming significance of the unconscious (in the Freudian sense of that word)." The truth is, he never had.

The development and ultimate souring of their personal relationship is, to a large extent, paralleled by citations Freud and Stekel made of each other's works, which we now briefly review. The biggest favor Freud did for Stekel was to write an introduction to *Nervous Anxiety States* (Stekel, 1908), although according to Stekel he had to make Freud rewrite it twice because the first two versions were rather "noncommittal and indifferent" (Stekel, 1950a, p. 119). Even in the published version, Freud's endorsement is still somewhat half-hearted: "If I am glad to take responsibility for this work [...] I think it is only right to declare explicitly that my direct influence upon the volume on nervous anxiety which lies before us has been a very slight one" (Freud, 1908d, pp. 250–251). More important, however, is that there are well

[9] Jones (1953–1957, Vol. III, p. 146) reported that Freud postponed publication of his book on fetishism until he found out "whether Stekel had touched on the solution he was now propounding in his book." See furthermore the *Rundbriefe* of the secret committee (Wittenberger & Tögel, 1999–2004), which provides evidence of the fact that Stekel's each and every move was anxiously watched and commented on.

[10] Jelliffe once wrote to Freud that he had the impression that Stekel longed for "a reconciliation with the master" (Burnham, 1983, p. 223). Jones (1953–1957, Vol. II, p. 102) reported on a letter by Stekel to Freud dated January 22, 1924, in which the former proposes that his and Freud's groups should cooperate "to fight against the common enemies"; Jones believed Freud never answered the letter.

[11] The "standing on the shoulders of a giant" claim surfaced in a *Rundbrief* of the secret committee, where Jones wrote (11 February, 1922): "Stekel's books are being translated in large numbers and he is enjoying a considerable vogue, especially in America but also here. People are especially enamored by his modesty! He has found the catchword that a dwarf on a giant's shoulders sees further than the giant." See for a further discussion of this metaphor, the chapter on marginal historiography in this volume (Wittenberger & Tögel, 1999–2004, Vol. III, p. 50).

over 50 identified allusions to, comments on, and direct references to his Viennese colleague in the *Standard Edition/Gesammelte Werke* volumes, of which almost half (approximately 20) make their appearance in the successive editions of *The Interpretation of Dreams* published between 1909 and 1925. Freud's very first reference to Stekel dates back to 1907, when Freud (1901, p. 120) added an example to the second edition of *The Psychopathology of Everyday Life* supplied by Stekel relating to a slip of the pen, the authenticity of which Freud was "willing to vouch for." Freud's final reference to Stekel is made in the *New Introductory Lectures* (1933, p. 144), where Freud counters the accusation brought against him of intolerance: "The only manifestation of this ugly characteristic has been precisely our parting from those who think differently from us. No harm has been done to them. [Apart] from one notable exception, it was they who excluded themselves." The editors sagely remarked in a footnote that this last allusion may possibly refer to Stekel.

Of Freud's references to Stekel prior to 1912, quite a few acknowledge his contribution to psychoanalysis, in particular his work on dream symbolism. For example, in the third edition of *The Interpretation of Dreams* (Freud, 1900), published in 1911, Stekel's work is mentioned in eight different places, most of which appear in the chapter on dream work, where a discussion of Stekel's dream symbols takes up a full page. More references to Stekel in this section appear (albeit in a more cautious tone) in the fourth edition of 1914, including Stekel's "largely confirmed" interpretation of numbers, bodily parts, and so on, as representing genitalia.

Although after 1912 Freud's references understandably become more cavilling, he was still able to write in the 1914 edition of *The Interpretation of Dreams* that "Stekel's writings, and in particular his *Die Sprache des Traumes* (1911c), *contain the fullest collection of interpretations of symbols* [italics added]" (Freud, 1900, p. 357), and he did not refrain from adding more examples from Stekel's work in the next edition of *The Interpretation of Dreams*, published that year.

References such as these are, we believe, somewhat undervalued in the historiography of psychoanalysis, for it is not often that Freud admits to having been under someone's influence. But how to trace this influence? One indication of an *indirect* influence of Stekel may be traced in Freud's references in *The Interpretation of Dreams* (1900) to Artemidoros von Dalis' *Oneirocritica* [Dream symbolism], perhaps the very first classical text on dream interpretation, written sometime in the second century AD. Oddly, there appears only one reference in the first editions of *The Interpretation of Dreams* to this text, and even then only in an abstract of a work by Gruppe. This made Wolfram Kurth (1951) wonder whether Freud had read the book at all. But in the 3rd (1911) and 4th (1914) editions several lengthy direct references to Artemidoros were added, using the 1881 translation by Krauss. Now, Friedrich Salomo Krauss (1859–1939), ethnographer and folklorist and editor of *Antropophyteia*, was an irregular visitor to the Vienna psychoanalytic society from 1910 on. No doubt this personal contact induced Freud to pick up his translation of Artemidoros. But he was not the only one to do so. Stekel's 1911 *Die Sprache des Traumes*, which appeared slightly earlier than Freud's 1911 edition of *The Interpretation of Dreams*, had abundant discussions of and quotations from this book. "How complete Artemidoros von Dalis" symbolism appears, whose book on dream symbolism even today is worth reading for psychoanalysts', Stekel (1922, p. 6) exclaimed. It seems that Stekel's enthusiasm for Artemidoros' dream interpretations stimulated Freud to adopt from him several Artemidorian-like sexual dream symbols, such as "left" for homosexuality, "right" for marriage and "luggage" for a load of sin (Freud, 1900, p. 358). But we should note the differences between Stekel and Freud in this respect too. The main claim of Artemedoros was that every element in the dream has symbolic meaning, and that by interpreting these meanings "objectively," it is possible to predict the future of the dreamer. Stekel used Artemidoros' procedure without any reservation, except that he did not aim to predict the patient's future but his past, so to speak. To Stekel, the dream symbols present a direct entrance into the unconscious, and his analyses of the dreams stop once its symbolic meanings are unravelled. To Freud, however, the interpretation of dream symbolism is only the first step into a long and painful process of self-awareness, by no means completed with a reading of the dream symbols.

The point of all this is to say that *influence* should be measured not only by looking at all the signs of B adopting a particular viewpoint under the suggestion of A, but also by looking at *parallel* processes taking place in A and B.

In *On the History of the Psychoanalytic Movement*, also published in 1914, Stekel is only briefly referred to as someone "who at first did such very credible work but afterward went totally astray" (Freud, 1914, p. 19ff). Freud's concern was exclusively with Adler and Jung; Stekel's defection was not dealt with at any length. However, in the sixth edition of *The Interpretation of Dreams*, published in 1925, Freud decided to devote a section on Stekel after all, explaining that "that writer, who has perhaps damaged psychoanalysis as much as he has benefited it, brought forward a large number of unsuspected translations of symbols; to begin with they were met with scepticism, but later they were generally confirmed and had to be accepted" (Freud, 1900, p. 350).

Although many of Freud's references to Stekel are thus perfectly open, if at the same time also somewhat ambiguous, many others are much more camouflaged. The editors of the *Standard Edition* were able to identify a good number of them, such as the one in the *New Introductory Lectures* mentioned above, but they missed or misattributed several others. We now briefly discuss one example and deal with another more extensively later on. In his study on Leonardo da Vinci, Freud (1910b) mentioned that the artist was left-handed and describes how this fact may be of importance, to which the editors added in a footnote with a reference to the *Three Essays on Sexuality* how this is "no doubt an allusion to the views of Fliess." However, in *Three Essays* (Freud, 1905) there is no connection between Fliess's notion of homosexuality and left-handedness to be found. In fact, it was Stekel who in a section on "right and left in dreams" in a contribution to the *Jahrbuch* (Stekel, 1908, p. 470) claimed that "homosexuality is associated in dreams with the left side of the body." He then proceeded to explain how "this would match an observation of Fliess', that so many homosexuals are left-handed." So, if there is an allusion to Fliess at all in this observation of Freud, it is one that comes about via Stekel. (Freud later referred to Stekel's left and right contribution in his 1914 edition of *The Interpretation of Dreams* [1900, pp. 357–358]; see Clark-Lowes, 2001, for two more hidden allusions elsewhere to Stekel.) We discuss later how one of these misattributed allusions to Stekel came to play an important role in the Freud–Stekel relationship, but first we examine how their reference policies with regard to one another changed.

It is true that although Freud would only slightly alter his references to Stekel ("my colleague Wilhelm Stekel" became "Wilhelm Stekel," "a true explanation" became a "further explanation," etc.), Stekel's revenge was more serious. He would withhold Freud's introduction to further editions of his *Nervous Anxiety States* and become openly critical toward Freud, although, in contrast to other "defectors," he continued to write books and articles as a psychoanalyst, some of which were even reviewed in psychoanalytic journals, although largely in a dismissive tone. Outside traditional or strictly Freudian circles (where he was considered a persona non grata), Stekel succeeded in generating quite a following, particularly in the United States, where together with Samuel Tannenbaum he presided over a short-lived journal. There he found in James van Teslaar a tireless translator and in Emil Gutheil a loyal supporter who published his memoirs after his death and who was also responsible for the editing of most of Stekel's posthumous English translations.

In 1926, Stekel published his version of his falling out with Freud (here published as chapter 8), stressing the scientific differences between them and Freud's intolerance for differing opinions as the basis for their break-up. However, Freud wrote:

> I contradict your often-repeated assertion that you were rejected by me on account of scientific differences. This sounds quite good in public, but it doesn't correspond to the truth. It was exclusively your personal qualities—usually described as character and behaviour—which made collaboration with you impossible for my friends and myself (Freud to Stekel, 13.1.1924).

This embittered Stekel: "I do not want to compare myself with Freud. I have never assumed this. All I want is justice. I have received much from Freud. But I gave him at least as much, if not more," he wrote in that same historical account.

One does not have to disagree with Freud if one neither believes that his personal aversion toward his former disciple is sufficient explanation for Stekel's downfall as an analyst nor that his initial sympathy for him is sufficient explanation for his initially becoming one. These personal motives certainly played a role, but other factors were equally important in this process.

Of Jensen and Other Poets

In the summer of 1906, Freud wrote an analysis of Wilhelm Jensen's novel *Gradiva*, which had been published 3 years earlier. The novel is about an archaeologist who falls in love with an antique sculpture of a young girl only to gradually realize that the sculpture represents a living girl with whom he had been in love since childhood. Repression of this erotic impulse drove him into a delusional and obsessive love from which he was freed only after being confronted with the object of his passion, the girl herself (who thereby acted as a therapist). Freud's analysis revealed how the novel's author unconsciously followed all the "classical rules of interpretation" in order to make the story work. The "delusions" described are so realistic in a psychiatric sense, wrote Freud, that one wonders if the author had knowledge of his *The Interpretation of Dreams*.

Freud published this study as the first volume in the series *Schriften zur angewandten Seelenkunde* [Writings on Applied Psychology]. That he wrote this particular study at this particular moment is no coincidence. According to Jones (1953–1957, Vol. II, p. 341), it was "Jung who drew his attention to the novel," and allegedly it was written "expressly to give him [Jung] pleasure." Both the first part of this claim, that it was Jung who brought *Gradiva* to Freud's attention, as well as the second part of it, that it was especially written to please Jung, are based on a small but significant mistake. We will show that it was not Jung but Stekel who brought the book to Freud's attention and that its purpose was not to please but to educate.

We first concentrate on what made Jones, and subsequently almost every historian, believe that it was Jung who brought this book to Freud's attention.[12] Jones's source is Jung himself, who apparently communicated this detail to him. Jones realized, however, that if Jung was not mistaken *Gradiva* must have been written before Jung and Freud actually met each other because their correspondence commenced in April 1906, and it would be some time before they met face to face; that is, if Freud wrote this

12 Every historian but one. The editors of the Freud–Jung correspondence were puzzled by this allusion to Jung and found no evidence of it, but Kerr (1994, p. 145) was the first (and, to our knowledge, the only) author who firmly denied the connection: "*Gradiva* was so important to Jung that it became part of the history of the work that it was written especially for him. It was not."

study to do Jung a favor it would be on account of an exchange of letters that had only just begun with a man whom he had not even met. As unlikely as it is, this is not impossible, and Jones was inclined to believe that it corresponded "with the high esteem in which Freud already held Jung at that time."

The reason why Jung claimed that this study was written to please him remains unexplained, other than by a misrecollection either on his or Jones's part. Jones's reason for holding this belief is, in our opinion, easier to understand: He misinterpreted two crucial references in the *Gradiva* study. The first occurs in the postscript to the second edition of Freud's study, in which he indicated that "a friend of mine has since then drawn attention to two other of the author's stories" wherein elements of the form and content of the *Gradiva* theme seem to be repeated, pointing also at "an intimate association in childhood of a brother-and-sister kind" (Freud, 1907a, p. 94). This friend is without any doubt Jung, because late in 1907 Jung wrote to Freud saying he found two other stories by Jensen "wonderful parallels of Gradiva. [...] *The problem is one of brother–sister love*" (Jung to Freud, 2 Nov. 1907). The second reference in the *Gradiva* study is less clear. Freud mentions "a group of men" who are interested in solving "the essential riddles of dreaming" and then continues to single out one among this group who had drawn his attention to Jensen's book (Freud, 1907a, p. 7 and 9). One must assume that Jones believed this person also to be Jung; he is, in any case, now identified as Jung in an editorial footnote (Jones, 1953–1957).

However, this person, "the one who, as I said at the beginning, was interested in the dreams of Gradiva," comes to the fore in a final section in the book when it is recounted how he "approached [Jensen] with the direct question whether he knew anything of such theories as these [i.e., psychoanalysis]. The author replied, as was to be expected, in the negative, and, indeed, somewhat brusquely [*unwirsch*]" (Freud, 1907a, p. 91). A confirmation of the fact that someone had asked Jensen about this matter is found in Jensen's letters to Freud, published in 1929, in which one can read how Jensen wrote to Freud on 13 May 1907: "That I gave a 'somewhat brusque' [*sogar etwas unwirsch*] answer to such a request I do not recall at all, and if this really happened I regret it and beg you to tell this gentleman: peccavi [I have sinned]" (Jensen, 1929, pp. 207–208).

From the Freud–Jung correspondence one learns that it is not merely extremely unlikely, but in fact impossible, that "this gentleman" refers to Jung, for a very simple reason: in his correspondence with Freud, Jung neither refers to a previous correspondence with Jensen nor betrays even the slightest knowledge of the novel's existence. So, when Jung read Freud's *Gradiva* study some time in May 1907 ("in one go"), he wrote to Freud to tell him how "magnificent" he thought it was. The question that comes to his mind is: "What does Jensen himself say about it?" (Jung to Freud, 24 May 1907). Freud then proceeded to inform Jung of the content of Jensen's letter to him but does not repeat his "peccavi," which he certainly would have, had it been addressed to Jung.[13]

So if it cannot be Jung, why must it be Stekel?

In a lengthy book on "The Dreams of Poets" (*Die Träume der Dichter*), to which we return shortly, Stekel (1912b) wrote how Freud,

> in an interesting essay, titled "Delusions and Dreams in Jensen's Gradiva," provided proof of the fact that Jensen's dreams, which he had fabricated at random, follow the laws of dream interpretation completely. I have asked Jensen whether he knew Freud's *The Interpretation of Dreams* and received a very sympathetic [*liebenswürdig*] reply that said he had no inkling of any dream interpretation whatsoever, and he had completely made up the story, just like that (Stekel, 1912b, p. 14).

Even though, typically perhaps, the "brusque reply" transformed into a "sympathetic" one, there can be no doubt that this is the inquiry Jensen mentioned in his letter to Freud and therefore that it is Stekel to whom Freud referred when he said that his attention had been drawn to the novel by someone "interested in solving the riddles of dreams."

The significance of this fact becomes clearer when one reads the passages in which these references to Stekel are made more carefully. In the very opening lines of his study, Freud relates:

> A group of men who regarded it as a settled fact that the essential riddles of dreaming have been solved by the efforts of the author of the present work found their curiosity aroused one day by the question of the class of dreams that have never been dreamt at all—dreams created by imaginative writers and ascribed to invented characters in the course of a story (Freud, 1907a, p. 7).

Of these men, one, who is now identified as Stekel, is singled out:

> It happened that in this group of men among whom the notion [of solving the riddles of dreams] first arose, there was one who recalled that in the work of fiction that had last caught his fancy there were several dreams which had, as it were, looked at him with familiar faces and invited him to attempt to apply to them the method of *The Interpretation of Dreams*. He confessed that the subject matter of the little book and the scene in which it was laid may no doubt have played the chief part in creating his enjoyment. [...] During the treatment of this genuinely poetic material the reader had been stirred by all kinds of thoughts akin to it and in harmony with it. The work was a short tale by Wilhelm Jensen (Freud, 1907a, pp. 9–10).

What is interesting in these descriptions is how this group of men in general, and "the one among them" in particular, are represented here. They are

[13] After completion of this chapter, we learned from P. Rudnytsky that he came to the same conclusion in a paper published in 1994 (in S.L. Gilman, et al., (Eds.), *Reading Freud's Reading*, 211–231); all priority for this discovery therefore must go to this author.

clearly students of Freud who share a special position vis-à-vis traditional scientists and who consider it a "settled fact that the essential riddles of dreaming have been solved" by the efforts of the author of *The Interpretation of Dreams*. They are, as followers of Freud, among the few who attempt to make use of the basic principles of psychoanalysis, but they are novices: Their interest in learning the trade is what characterizes them, hence their anonymity. The "one" singled out, however, is clearly no anonymous student; he stands closer to Freud in the story than the others. In fact, he is portrayed from the "inside," from the point of view not of Freud but of himself, a rare but significant occasion in Freud's work of the use of the technique of "imaginative writing" that is so characteristic of the psychological novel (Freud, 1908c, p. 150).

Knowing now that it was Stekel to whom Freud referred in such a peculiar way, one can come to a fuller understanding of this particular excerpt and even, we believe, to a different understanding of the text itself. We begin with two observations. First, it was Stekel who had discussed with Freud the idea that this novelist possessed an intuitive understanding of psychoanalysis; second, Stekel was the only member of the Vienna society who referred to Freud by his surname instead of as "Herr Professor" (Jones, 1953–1957, Vol. II, p. 8). This alone would make him something of a silent confidante in the *Gradiva* issue. Remember, however, that Stekel was no docile follower. He was a rebellious student, and in more than one way he occupied an exceptional position within the Vienna society because he endorsed some of basic psychoanalytic assumptions but also, and most explicitly, disagreed with Freud in other, sometimes fundamental ways. In psychoanalytic vocabulary, he showed strong signs of *resistance*.

This leads to the supposition that by referring to Stekel in this way, by referring to this mixture of confidence, resistance, and loyalty, Freud imposed on him the position of a patient. The patient/analysant is an antagonist who at once challenges the position of the analyst and at the same time puts himself in the position to be taught and in doing so commits himself to a bond of transference with the speaker. So the patient (Stekel) comes to embody an aspect of the ideal public that Freud was apparently looking for: the perfect antipode in his narrative, a point of reference that he needed to develop his own point of view. Creating such an antipode in his discourse was not new; he had done that before and would do so again afterward.[14] However, of course he imposed the patient position on his audience predominantly whenever he expected resistance. However, in contrast with this particular discourse, he would not impose this position

[14] In his lectures at Clark University, Freud (1910a) explicitly placed his audience in the patient position; in *The Question of Lay-Analysis* (Freud, 1926) he used the "impartial person" to impose on the audience the position of the patient (see Bos, 2001; cf. *Future of An Illusion* and *Civilizations and Its Discontents*).

on a specific person, but rather on the collective as a whole, or on some abstract or idealized character. By picking a very real person (albeit carefully disguised), by addressing one of his students directly, he put himself in a very specific position toward his readers, a position that constitutes the way one reads the book.

In what sense is the position of the listener/reader constitutive of the way one reads this narrative? One of his readers, Ernerst Jones, thought it a "charming book," "written with such delicacy and beauty of language as to rank high and to compel admiration for its literary qualities alone" (Jones, 1953–1957, Vol. II, p. 341). The *Gradiva* study admittedly is a compelling, even moving narrative, written almost as if the author had been carried away by his own enthusiasm. However, Jones was mistaken: It was not written to please anyone; its aim was to educate. Underneath its charm lays a very explicit tutorial discourse. The *Gradiva* study is first and foremost a textbook in disguise, a lesson in applied psychoanalysis. We now develop this point a little further.

Readers should note how Freud, after he announces his intentions to give a "brief summary" of the contents of the novel (1907a, p. 10), in fact devotes 15 full pages (one-third of the entire book) to the story and then asks himself (in wonder!) how it has come about "that we have been led into dissecting the whole story and examining the mental processes" (ibid, p. 41). Well, he explains, "this has not in fact been an unnecessary piece of work; it was an essential preliminary," and the reader understands that this is not an ad hoc excuse for wasting so much of his or her time; this is Step 1 of an exercise in "applied psychoanalysis." Read this way, the *Gradiva* study reveals many hidden "directives," both explicit and implicit. Among the more explicit instructions are references to *The Interpretation of Dreams* that take the form of "rules."

In trying to unravel one Norbert Hanold's dreams (the story's hero), Freud wrote:

> This dream is not self-explanatory and we must resolve to borrow from my "Interpretation of Dreams" and apply to the present example a few of the rules to be found in it for the resolution of dreams. *One of these rules is to the effect that a dream is invariably related to the events of the day before the dream.* [...] Another rule tells us that, *if a belief in the reality of the dream-images persists unusually long, so that one cannot tear oneself out of the dream, this is not a mistaken judgment provoked by the vividness of the dream-images, but is a psychical act on its own: it is an assurance,* relating to the content of the dream that something is really as one has dreamt, it is right to have faith in this assurance (Freud, 1907a, p. 57, emphasis added).

Elsewhere, Freud wrote: "There is a rule in interpreting dreams, which runs as follows: 'A speech heard in a dream is always derived from one that has been heard or made by the dreamer in waking life'" (Freud, 1907a, p. 74).

This exhibition of law-like "rules" is, in our opinion, no longer part of an explanation of how Freud came to analyse this particular dream; it is part of a general exhibition of psychoanalytic technique. They bear the characteristics of a teacher's instruction, complete with a teacher's position and corresponding tone of voice (decisive yet patient). Some other passages equally bear these characteristics, and although they are not referred to as "rules," they certainly bear that mark. Two examples are (a) "If ridicule, derision, or embittered contradiction occurs in the dream-thoughts, this is expressed by the manifest dream being given a senseless form, by absurdity in the dream" (Freud, 1907a, p. 82) and (b) "We are warned by that theory [psychoanalysis] not to fall into the error of tracing the anxiety that may be felt in a dream to the content of the dream, and not to treat the content of the dream as though it were the content of an idea occurring in waking life" (Freud, 1907a, p. 60).

There are more implicit instructions as well in which pedagogy is interwoven with analysis. In explaining the heroine's behavior, for example, Freud drew a parallel to the analytical situation:

> We are beginning to understand now, and to feel some hope. If the young lady in whose form Gradiva had come to life again accepted Hanold's delusion so fully, she was probably doing so in order to set him free from it. There was no other way of doing so; to contradict it would have put an end to any such possibility. *Even the serious treatment of a real case of illness of the kind could proceed in no other way than to begin by taking up the same ground as the delusional structure and then investigating it as completely as possible* [emphasis added] (Freud, 1907a, pp. 21–22).

"Delusions and Dreams in Jensen's *Gradiva*" is thus presented to us as an *exemplary analysis*. It is analysis in the making: we follow each of Freud's steps while he analyzes. We gradually discover the meaning of certain dream symbols, we find out how specific, seemingly unrelated events or utterances are in fact related, and in this process we are taught to pay attention to every little detail, not to jump to conclusions, to question and requestion ourselves, and, above all, we are taught how to apply the "rules of interpretation."

If this reading of Freud's analysis appears to be forced, we might respond that Freud himself was also unsure about it, believing that "the problem of the relationship between the poet's creations and his personal life [...] has remained unsolved" in this book (*Minutes*, Vol. I, p. 246).[15] More important, however, Gilman (1993, p. 146) convincingly showed that it was Freud who read into Jensen's novel more than was there, or rather that *his* psychoanalytic reading of it is somewhat forced and does not do

[15] It is of interest here to note that Freud made this comment to the Vienna Psychoanalytic Society after a presentation by Stekel.

justice to the original because the main theme of the novel is not about the genesis and cure of a delusion at all but about something entirely different, something that "Freud does not notice"; it is about racial differences. So the novel is no more than an excuse, a pretext, for Freud to present his analytic technique.

If the *Gradiva* study was indeed designed as a course in psychoanalysis, are there any indications that Freud's students consumed this work as such? Recall that it is in his poet's dreams book that Stekel revealed he had written to ask Jensen whether he had ever read Freud's *The Interpretation of Dreams*. He was prompted to ask him this, Stekel said, because "the dreams he had constructed [in his novel] completely follow the rules of dream interpretation" (Stekel, 1912b, p. 14). We now pick up on the argument Stekel began to make here. If the dreams of poets, such as Jensen or Goethe (either in fabricated form as a piece of fiction, or actually dreamed) are not only very similar in content and structure, but also seem to follow the rules of dream interpretation very closely, then there must be a connection between poetry and neurosis. This was the main claim of Stekel's (1909b) book "Poetry and Neurosis" from which he subsequently deduced the following questions and suppositions:

> Is it true [...] that poets dream more expressively and with more fantasy? Can we perhaps find in their dreams a path to understand poets? Because once we know how to decipher the language of dreams, we have in the dreams of the poet objective material that will reveal to us his innermost secret and hidden thoughts. Such a study seemed to me both exciting and rewarding (Stekel, 1912b, p. 16).

Having convinced himself that it *is* rewarding and exciting to investigate the dreams of poets, Stekel subsequently described how he designed a questionnaire that he sent out to an unknown (but presumably large) number of Austrian and German poets and writers. He received no fewer than 38 replies, including from some of the most renowned writers of his day, such as the Nobel Prize laureate Paul Heyse (1839–1914) and the famous author of wild west stories Karl May (1842–1912). Stekel furthermore searched these authors' writings for dreams and in this way collected a large amount of data, which comprised the main body of his book.

This approach, of collecting case material from poets and writers in order to study the connection between dream and creativity (or "poetic production," [*dichterische Produktion*]), goes directly back to a suggestion Freud made in his *Gradiva* study. In fact, wrote Freud, after having mentioned how the problem of poetic production is related to the investigation of dreams, there are only two possible lines of research:

> One would be to enter deeply into a particular case, into the dream-creations of one author in one of his works. *The other would be to bring together and contrast all the examples that could be found of the use of dreams*

> *in the works of different authors* [italics added]. The second method would seem to be far the more effective and perhaps the only justifiable one, for it frees us at once from the difficulties involved in adopting the artificial concept of "writers" as a class. On investigating this class falls apart into individual writers of the most various worth—among them some whom we are accustomed to honor as the deepest observers of the human mind (Freud, 1907a, p. 9).

Of Freud's two options, Stekel chose the second, the one that deemed Freud the more appropriate, indeed perhaps the "only justifiable one"—the royal road into the unconscious, so to speak. With no further explanation, Freud nonetheless chose the first route. We now know, however, that he never intended to investigate the problem of "poetic production": his intention was to use the *Gradiva* analysis for a different, tutorial purpose.

In a way, Stekel too had a hidden agenda. His book starts off as an attempt to "solve the riddles of the dream" using two explicit Freudian assumptions: (a) that all dreams are wish fulfilments and (b) that each dream contains a sexual motive. Had he stuck to these two premises, his study could have been acceptable, even valuable, to Freud, despite Stekel's insistence on "bipolar" forces and his infrequent acknowledgement of Freud. However, Stekel's desire to be original was more powerful, and he anticipated that, standing on the shoulders of a giant, he could see further than the master. He thus expanded his approach. The questionnaire he sent out consisted of six questions: (a) Do you have typical (reoccurring) dreams? (b) Can you tell me about a dream which made a deep impression upon you? (c) Do you daydream? (d) Do your dreams ever involve criminality? (e) Are your dreams ordinary or fantastic? (f) Do you make use of your dreams for poetic purposes? (Stekel, 1912, p. 36).

The first three as well as the last question derived from Freud; Questions d and e, however, clearly did not. They came from Stekel's own "discovery": the "universal criminal" aspect in dreams [*universell kriminell*]:

> It is to my own merit—I dare say—that I have discovered the criminal aspects in the dynamics of the neuroses. Also I have revealed the secret symbolic language of criminality in dreams. I can, therefore, based on my experience in translating the language of dreams, disclose the secret criminal drives in dreams without the aid of psychoanalysis, and to a large degree of probability (Stekel, 1912b, pp. 35–36).

Having collected dreams and other relevant material Stekel, instead of applying the "rules of interpretation" to the dreams of poets and writers, drifted off into explorations and interpretations of a whole variety of "criminal" aspects, reaching a conclusion that was not even remotely Freudian but in fact more Adlerian, namely, that poets, writers, criminals, and neurotics all suffer from a "delusion of grandeur": they all share a "secret belief in their own great historic mission" (Stekel, 1912b, p. 251).

If, from Freud's point of view, Stekel's book on poets' dreams demonstrates anything, it is that his lessons were wasted on Stekel. Still, they were clearly *his* lessons wasted on him. In the entire design of the book, in the formulation of law-like principles ("the law of bipolarity," "the child is a universal criminal"), and even in the tone of voice, which is oftentimes forced, one recognizes not the hand of the master but an attempt to imitate it.

In the next section we explore the master–pupil relationship further.

"I Know That I Have a Destiny to Fulfil. I Cannot Escape It and I Need Not Move Toward it"

In all of the four volumes of the *Minutes* (Nunberg & Federn, 1962–1975), no fewer than 16 of the 141 regular meetings of the society (up to November 6, 1912, when Stekel resigned)—that is, over 11%—were devoted to presentations and papers by Stekel.[16] These documents also demonstrate that no participant of the meetings of the Vienna society contributed more to discussions, including Freud himself, than Stekel. In addition to attending the most sessions, there is hardly a meeting during which he did not participate. The *Minutes* have been called a testimony to the early history of psychoanalysis and to the way Freud taught his first students, inasmuch as they show how "all the functions of present-day psychoanalytic societies and institutes" were encompassed in these meetings (Solomon, 1969). And although it would be an exaggeration to say that the *Minutes* are a testimony to one teacher–student relationship in particular, namely, that of Freud and Stekel, it is certainly the one that dominates them. We explore some aspects of this relationship in detail here, focusing on two issues (a) Stekel's presentations from 1907 to 1910 to the society and (b) Freud's response to them.

Because almost all of Stekel's presentations to the Vienna society were eventually published as chapters in book form, we have grouped his papers into categories related to these publications. Accordingly, we propose to differentiate between contributions on the problems of (a) anxiety neurosis (*Nervöse Angstzustände*, 1908), (b) poetry and artistic writing (*Dichtung und Neurose*, 1909b), (c) doubt (*Der Zweifel*, 1912a), (d) symbolism in dreams and elsewhere (*Die Sprache des Traumes*, 1911c), and (e) masturbation (*Onanie-debatte*,

[16] In comparison, only Sadger (with 14 presentations) and Adler (11) come close to Stekel's 16. Freud presented 9 papers during this period, which was approximately the average for a frequent speaker (but many presented no more than 1 paper each year). We have excluded from our count 35 meetings on general themes (case reports, general discussions, business meetings).

in Dattner, 1965).[17] Setting aside the last group for the moment, we examine Stekel's position vis-à-vis Freud and explore whether their respective positions changed over the years.

We have already observed that it is remarkable that in almost all of the earlier presentations Stekel already and most explicitly digresses from Freud in important respects, most notably on the question of the etiology of anxiety neurosis. He is recorded to have advanced, for example, the following thesis:

> [That] all anxiety is fear of death. [...] Freud has traced anxiety neurosis back to coitus interruptus: he sees the origin of anxiety in the deflection of the somatic sexual excitement from the psychic sphere. Stekel stresses that, in contrast to this opinion, he must adhere to the conviction that also in anxiety neurosis—as in any other neurosis—the psychic conflict is essential (*Minutes*, Vol. I, p. 175).

Note that although this thesis is presented "in contrast" to Freud's opinion, it is not presented in opposition to it, as Adler eventually would do with his ideas. In fact, whenever Stekel criticized Freud, he presented his critique as in line with or following up on some basic assumption of Freud and never in direct disagreement with him. More frequently than any other member of the Vienna society, Stekel continually sought to expand psychoanalysis. Although he was often unsuccessful, his attempts to push to the limits psychoanalytic knowledge seemed to have annoyed Freud only occasionally; these attempts seemed to have been experienced by Freud as challenges, or at least as opportunities to explore terrain that had not yet been cultivated.

It is clear, then, that when Stekel attempted to define anxiety neurosis in terms of conflict, Freud chose not to dismiss the idea but to oppose it. He was critical toward Stekel, to be sure, but he also suggested some interesting parallels to follow up. He proposed that if anxiety neurosis is the somatic equivalent of hysteria, then there must also be a somatic equivalent to paranoia, namely, hypochondria (*Minutes*, Vol. I, p. 110). This was a direct hint to Stekel on how to investigate anxiety, which Stekel followed shortly afterward in a presentation on "the somatic equivalents of anxiety and their differential diagnosis" (*Minutes*, Vol. I, p. 204). Unfortunately, by presenting somatic *symptoms* in anxiety (of the heart, bladder, skin, stomach, etc.), he misunderstood the point, said Freud: he was not concerned with anxiety *equivalents* but

[17] The presentations on *anxiety neurosis* cover the protocols of Meetings 12, 21, 25, 30, 31, 35, and 50 (roughly 1907 to summer 1908); those on *poetry* cover the protocols of Meetings 54b, 64, and 74 (winter 1908–1909), those on *dream symbolism* cover the protocols of Meetings 108, 117, and 118 (spring 1910), those on *masturbation* cover the protocols of Meetings 109, 110, 111, 151, 153, 155, 159, 161, 164, and 170 (summer 1910, winter 1911 to summer 1912); and the one on *doubt* is in the protocol for Meeting 94 (early 1910).

rather with equivalents of anxiety *attacks* (*Minutes*, Vol. I, p. 209). It was not the physical but the psychological processes that interested him, a point on which Stekel and he would never find common ground.

Freud thus frequently had to "correct" Stekel, but he would also always use the opportunity to indicate more Freudian avenues of research. Therefore, when Stekel misdiagnosed a case of hysteria as a neurosis, Freud lectured him on "anxiety as a central problem in the theory of neurosis," again using Stekel as his ideal antipode, who is at once both proponent *and* opponent.

Anxiety is a phenomenon composed of physiological manifestations plus psychic sensations, but one cannot determine its origin by looking at this. Therefore, it is not valid to maintain that all anxiety must be derived from the psyche; with the separation of ordinary (life) anxiety from neurosis anxiety, it becomes just as untenable to maintain that every anxiety is fear of death (*Minutes*, Vol. I, p. 181).

On occasion, Stekel was more successful, such as when he began to make experimental use of Jung's association technique in therapy. His method consisted of prompting a patient with stimulus words to which he or she was to reply with a free association ("dog–fidelity," "road–width," "flower–red," etc.). He would then have the patient explain the free association back to him, again via free associations: "the dog is loyal—the patient leads a dog's life" (Stekel, 1908, p. 411). Freud at first believed these "technical innovations" to be "useful" (*Minutes*, Vol. I, p. 180), although he later became more sceptical:

> Associations as well as free thoughts yield much chaff. It just does not happen that at the very beginning of a treatment nothing comes to the patient's mind (so that one is forced to use the association technique of Jung) if one encourages him first to tell his story (*Minutes*, Vol. I, p. 250).[18]

A rather free or "wild" interpretation of psychoanalysis met with criticism and scepticism but also with encouragement: The Freud–Stekel relationship was, at least until about 1909, very antagonistic, but it was also somehow productive for both parties, even if neither felt particularly good about it.[19]

Did this customary antagonism characterize their entire relationship? No. There is evidence that by the end of 1908 both Stekel and Freud began

[18] It is an ironic twist of fate that precisely this type of association technique has become prototypical in the layman's eyes for the analytical situation. A beautiful parody of it is given in "The Singing Detective," when the interaction between the hero of the story and the analyst results in the following association-dialogue: "woman–fuck," "fuck–dirt," "dirt–death."

[19] Perhaps there was a certain sadomasochistic element in their relationship, considering that Freud never tired of criticizing Stekel, while Stekel allowed himself to be put in his place time and again—until at one point he is recorded to have compared "the neurosis of the teacher who rejects whatever is new, with the neurosis of the pupil for which he himself has so frequently been reproached" (*Minutes*, Vol. I, p. 253).

to realize that their working relationship was, to say the least, strained, if not exhausted. After a presentation by Stekel on the poet Grillparzer, Freud remarked that he feared that "the criticism directed against Stekel will make no impression on him, although if he did take the objections into account and eliminate certain exaggerations, the work would have some merit" (*Minutes*, Vol. II, p. 9), to which Stekel somewhat cynically replied that his paper would have fulfilled its purpose "if it accomplished no more than to serve as the occasion for the Professor's reply" (*Minutes*, Vol. II, p. 12).

Stekel's presentations on poetry and neurosis of late 1908 and early 1909 came much closer to a "classical" Freudian approach but were nevertheless criticized by Freud in strong terms. His analyses, Freud said, will no doubt impress the general public but not Freud and his followers: "It is too general on the one hand, too superficial on the other. Rank has a deeper conception of the topic" (*Minutes*, Vol. II, p. 103).

The closer Stekel sought to come to Freud, the more severe and hostile the critique became. A "contribution to the psychology of doubt," presented a year later and expressly written as a modest attempt to follow up some ideas Freud had published in his Ratman case, has abundant references to Freud, but it received a terrible reception. Freud sarcastically remarked that he published his case with "the purpose of stimulating others to collaborate" and was glad that "at least in this sense the stimulus was successful" (*Minutes*, Vol. II, p. 400), but he thought Stekel's theoretical section was "weak and blurred," that he had "overlooked the main characteristics of doubt," and that he "mixed up cases of anxiety hysteria and obsessional neurosis." He now even rejected Stekel's favorite term *bipolarity* as an unhappy choice: "It no longer describes something new" (*Minutes*, Vol, pp. 401–402). Desperate, or at least weary, Stekel replied that he was "already accustomed to the fact that everything new he offers encounters great resistance at first, only to be accepted in many instances after publication—above all by Professor Freud" (*Minutes*, Vol. II, p. 403). Freud refused to permit the paper to be published in the *Zentralblatt*—the first time he used his veto against him, Stekel later claimed (1926, p. 558).

Two papers on symbolism (in dreams and in the choice of a profession), published in the *Zentralblatt*, as well as a discussion on Stekel's *The Language of Dreams* (1911c), mark the last phase of their relationship. Freud's comments on these papers were now more sparing than on Stekel's previous presentations at the society but at the same time less encouraging. Of the paper on dream symbolism Freud said only that its topic "is a good example of the questions that are posed by those to whom our psychological points of view are alien" (*Minutes*, Vol. II, p. 537). In the discussions on Stekel's paper on the choice of a profession, Freud was for the first time not even present. Freud contributed a *written* review to discussions on Stekel's *The Language of Dreams,* which was, as noted earlier, so critical that Bleuler refused to print it in the *Jahrbuch*. In the *Minutes*, Freud is recorded to have said on occasion that "Stekel did not know how to keep himself within the limits. Not all dreams require the application of symbolism; many dreams can be resolved with the help of only a modest measure of symbolism" (*Minutes*, Vol. III, p. 237).

By 1911, Freud was clearly withdrawing, at least as far as Stekel was concerned, from his role as a teacher. His critique became less constructive and more outspokenly scathing of those elements that did not fit within his own framework. In his critical review of *The Language of Dreams* he is merely an opponent, no longer his teacher. It is interesting to note, however, that this was also the book to which Freud would most often refer, and in quite a laudatory manner. In the last section of this chapter we seek to answer this contradiction.

We now come to Stekel's last contributions to the society: those on the problem of masturbation. The historical record typically portrays Freud as an opponent, and Stekel as a proponent, of masturbation. In the next chapter we review the literature on this subject and trace the development of and the culminating point in the discussions between Freud and Stekel in the Vienna Psychoanalytic Society to a particular "shared interest." There we propose the hypothesis that, thanks to his brief therapy with Freud, Stekel realized that his "sexual problems" were related to feelings of guilt resulting from masturbatory practices and that he had to let go of these fears. Thereafter he became an advocate of its practice, while Freud, thanks to Stekel, realized he needed to maintain the view that masturbation is harmful in order to uphold his theory of anxiety neurosis, and he consequently opposed its practice.

These all-too-personal motives were activated again, we believe, in 1910 when Stekel began to propound in public his thesis of the harmlessness of masturbation, forcing Freud to organize a discussion on masturbation in the Vienna society so as to make it clear for all that Stekel's dissenting ideas no longer belonged to the psychoanalytic community. The *Zentralblatt* affair was only than a pretext to get rid of him, and this confrontation took place almost simultaneously with the publication of the "*Onanie-debatte*" in late 1912.

This hypothesis partly supports Stekel's long-held position that his dissident views on masturbation were the actual cause of his being expelled from the movement (see chapter 8). It is certainly true that his opposition to Freud in this matter contributed in no small part to his downfall, but we hesitate to emphasize this personal antagonism. Stekel had been a "dissident" from the very beginning, and his antagonism with Freud was in fact a constructive rather than a destructive force; that is, this was true as long as a clear master–pupil relationship was maintained. Once that relationship changed, Stekel's position became endangered. We therefore submit that conflict in the masturbation debates did not cause the break but was rather a result of it.

What are the points of contention between Freud and Stekel on this issue? According to Freud, the masturbatory phase is first an individual developmental phase. On various occasions he explained how, from pure (bodily) autoerotic sensations, individuals learn at a young age to obtain pleasure by merging wishful fantasies with self-gratification, resulting in masturbation. This practice is given up once the subject learns to attach his libido to objects outside his own body (Freud, 1905, 1908a). This is, strictly speaking, a normal—indeed, almost necessary—developmental phase, which "scarcely a single individual escapes" (Freud, 1905, p. 188). Masturbation becomes harmful in a bodily sense if the individual is unable to give up this practice in later life (or returns to it later on) and consequently may develop neurasthenia. Freud referred to this as the *toxic component* in masturbation, admitting though that the exact mechanism is "completely unknown to us at the present time" (*Minutes*, Vol. IV, p. 95). It is also harmful in a more important, psychological sense, because the individual does not learn to separate wishful fantasies from pleasure, allowing fixation to take place (*Minutes*, Vol. IV, p. 94). The return to a narcissistic stage surely has detrimental effects on the character, he hypothesized.

Stekel, on the other hand, had always maintained that it is not masturbation that leads to illness, but the other way around: repression of the urge to masturbate is harmful because it leads to feelings of guilt. These conflicting feelings (pleasure vs. guilt) result in anxiety neurosis. At least as early as 1908, Stekel emphasized this link, and right from the start he bore the brunt of Freud's critique. By 1910, Stekel flatly denied that masturbation engenders neurasthenia (*Minutes*, Vol. III, p. 338), but only in 1912 did he go as far as to deny the existence of neurasthenia altogether (*Minutes*, Vol. IV, p. 37), at which point he found himself in direct opposition to Freud, whom he accused of "drawing an incorrect conclusion in thinking that if infantile sexuality is harmful, masturbation too, must be. [...] It is not infantile sexuality that is harmful, but our attitude to it" (*Minutes*, Vol. IV, p. 42). In his published contribution to the "*Onanie-debatte*" Stekel (In Dattner, 1965, p. 34) differentiated between a wide variety of types of masturbation—conscious and unconscious, with and without "friction," with and without fantasy—and this extension of the concept of masturbation enabled him to claim that "everybody masturbates" (whether he knows it or not). Finally, Stekel went on to show through a case history presentation how masturbation actually could relieve a person from neurotic symptoms, rather than the other way around.

Several authors have pointed out that in retrospect it appears difficult to escape the conclusion that Stekel was right and Freud was wrong on this, because the "toxic component" in masturbation is a myth: There are no physical harmful effects to it. Indeed, masturbation has since become a much more accepted sexual practice, whereas neurasthenia as a disease disappeared from the stage altogether. However, it is not that simple. In his brilliant new book on solitary sex, Thomas Laqueur (2003) showed that

Freud played a key role in transforming the problem of masturbation from a purely medical issue into a more psychological one, even if he did not succeed in leaving behind the ethical issue. Although in the late 1880s, Krafft-Ebing (one of Stekel's mentors) still firmly believed that masturbation caused insanity, by the turn of the century this view slowly gave way, but it was not until Freud that the public view on onanism would change. Masturbation, Laqueur (p. 381 ff) wrote, "allowed Freud to connect the unfocused sexual energy—the autoerotic stage, as he later called it—with fantasy and the sort of inner wishing represented by later Oedipal desires." Giving up masturbation is a "cardinal sublimation" demanded by society, Freud thought; clinging on to it makes the sexual instinct uncontrollable and unserviceable to civilization. So it is not so much the physical act as such that is harmful, it is the fact that in this act fantasy finds no proper object that makes it a dangerous practice. From this perspective, Stekel's views are not in opposition to Freud: they are a perversion of it, a dangerous simplification, overemphasizing the moral side of the problem while denying society's pressure.

In the concluding remarks to the "Debate on Onanism," Freud (1912b) devoted almost his entire contribution to a refutation of Stekel's views, repeating arguments he had used many times, pointing out, for example, how Stekel "overstretched psychogenicity" in the etiology of neurosis (p. 246) and underlining once again how Stekel's counterarguments failed to meet the point. It is interesting to note, however, that Freud's refutation is presented here in a historical perspective. The antagonism between himself and Stekel is no longer a current opposition but a *historical* opposition, presented to the reader as a conflict that was bound to happen:

> Stekel shares with many nonpsychoanalytic writers an inclination to reject the morphological differentiation which we have made within the jumble of the neuroses and to lump them all together under one heading—under psychasthenia, perhaps. We have often contradicted him on this, and have held fast to our expectation that the morphologico-clinical differences will prove valuable as indications that have not yet been understood of essentially distinct processes (Freud, 1912b, pp. 248–249).

Stekel had become an outsider, a "nonpsychoanalytic writer," whose point of view is, admittedly, "courageous" and who has made "correct comments to us on the question" but who is definitely no longer part of that collective; the break between Stekel and Freud is not developing; it is already fact.

In an interesting, if often neglected, section from this same contribution, Freud reflected on how "clinging ... to his own ideas" should be understood in the light of his breaking away from former students. It is almost as if he personally addressed Stekel here, echoing the latter's phrase "great historic mission" of the poet when pointing out the inevitability of fate—for even if it turns out that he was wrong, Freud wrote, he could not have acted otherwise. He is moving toward a certain destiny in which Stekel no longer plays a part:

> If in the end I am convicted of being in error over the theoretical problem of the "actual neuroses," I shall be able to console myself with the advance in our knowledge, which must disregard the opinions of an individual. You may then ask why, since I have such a laudable appreciation of the limitations of my own infallibility, I do not at once give in to these new suggestions but prefer to reenact the familiar comedy of an old man clinging to his onions. My reply is that I do not yet see evidence to induce me to give in. In early days I made a number of alterations in my views and did not conceal them from the public. I was reproached on my account of these changes, just as today I am reproached for my conservativeness. Not that I should be intimidated by the one reproach or the other. But I know that I have a destiny to fulfil. I cannot escape it and I need not move toward it. I shall await it, and in the meantime I shall behave toward our science as earlier experiences has taught me (Freud, 1912b, p. 250).

The Dialogical Freud and the Language of Psychoanalysis

With their respective fates sealed, and Stekel effectively expelled from the psychoanalytic movement even before he refused to "unconditionally obey" Freud on the *Zentralblatt* issue, we have returned to the beginning of our inquiry. We now synopsize this study's findings on what perhaps could be called the *dialectics of an antagonistic interaction.* We begin by summarizing the main characteristics of this relationship.

Firstly, it is important to observe that the rapport between Stekel and Freud was, from the very beginning, tense, colored by feelings of ambivalence and doubt on both sides, and that these emotions arose long before, and persisted long after, their break. To Freud, Stekel never was, nor did he ever become, a "dissident"; he had always been, and remained, in a sense, a source of irritation as well as inspiration. As for Stekel, Freud had always been and would always remain a father–teacher rival before whom he wished to excel but whom he also wanted to surpass.

Secondly, these antagonistic feelings resulted, to a certain extent, in a constructive rather than a destructive dialogue. More importantly, these feelings prompted them to exceed their respective roles as student and teacher. Stekel tried to become Freud's *ultimate* student—not a mere follower but a creative, original scholar, a productive, imaginative, and above all rebellious advocate of psychoanalysis. One might even conjecture that many of Stekel's ideas might never have been formulated had they not been articulated in an explicitly contrasting (and later opposing) manner and that these ideas bear the mark of their dialogical origins. Groenendijk (2001, p. 150 footnote) noted how identification and differentiation went hand in hand in their relationship, underlining the very idea of bipolarity, which after all directly suggests a struggle between two antipodes. Furthermore, their respective viewpoints on the problem of masturbation need to be understood in the light of this personal relationship; one might even go so far as to suggest that Stekel's masturbation thesis bears a reflection of his antagonistic relationship with Freud, inasmuch as that the former, given Freud's aloofness, was left no alternative than a form of self-gratification with a "fantasized other." Lastly, Stekel's complaint that the Freudians adopted his ideas without proper acknowledgement should, viewed in this light, perhaps be understood not so much as a priority problem but as a problem of discontinued dialogue; once their interaction stopped, Stekel lost a willing ear and was left with a patient.

Dialogue with Stekel had also proven useful to Freud and confronted him with problems of a different kind once it discontinued. Mahony (1987, p. 60) observed how Freud, particularly in these early years, was keen on establishing direct audience contact but did "not use a full blown dialogue format to achieve the dialogue's feeling of immediacy." Instead, he hinted at it indirectly and through "manipulation of personal pronouns." We have shown how Freud, using this technique, implicitly moved Stekel into the discursive position of a patient by hinting at his deviant viewpoints on the one hand and articulating the bond of confidence on the other. It enabled Freud to open up a tutorial position in his discourse from which he could teach. Once their interaction stopped, he lost an unwilling but equally useful ear.

The question has been raised as to whether the Vienna society was not, in fact, a "therapeutic community" (Leupold-Löwenthal, 1980), and the answer is obviously, no, not in a literal sense. Inasmuch as this community allowed itself to be relegated to the position of pupil to the mentor, however, elements of the classic patient–analyst relationship were imposed. This explains an initial difficulty we had with one of the societies' conversational rules: Each member was obliged to participate in the discussion regardless of whether he had anything to contribute. The parallel with the "first rule" of therapy—the obligation to speak during analysis—is striking from this perspective. At a time before didactic analysis, the weekly meetings of the Vienna society came closest to a training analysis, and Freud, at least, seemed to have experimentally used an instructional type of analytic intervention when elements of transference and resistance came to the fore in the discussions. Perhaps in this respect, part of Stekel's rebellious conduct ought to be considered to be transference neurosis. Even if this is so, we propose that an analytic interpretation of the individual psychodynamic factors, though helpful in explaining struggle, friction, or even resistance at an interpersonal level, leaves open the question of the interpersonal dynamics, more specifically, the problem of change in a relationship.

We have suggested looking at the interpersonal discursive aspects to discover what made and unmade Stekel as an analyst. Earlier we proposed that what constituted the Freud–Stekel relationship was a dialogue based on mutual antagonism, framed within an analyst–patient setting. However, if the making of Stekel as an analyst was based on this unusual teacher–student dynamic, then the "unmaking" of Stekel must be attributable to a change in this dynamic force. Therefore, we suggest we look for these forces at a discursive level, that is, at the junction between the language of psychoanalysis and the social conditions within which its practice was situated.

In the crucial years between 1902 and 1910 psychoanalytic language changed as a result of organizational developments: the small, local, and loosely organized Viennese group (whose primary interaction consisted of face-to-face meetings) evolved into the large, formally organized international society with three official journals (the *Schriften* series, the *Jahrbuch,* and the *Zentralblatt*). These changes in organization are related to what Thompson (1995, p. 134) called "the transformation of visibility." The fact that organizational developments called for new forms of communication also implied new, more visible forms of control, as when face-to-face interaction is replaced by quasimediated interaction through use of written (printed) media, "those who exercise power, rather than those over whom power is exercised, are now subjected to a certain kind of visibility" (ibidem). In other words, if at one point Freud could single-handedly manage his authority in dialogical interactions with his followers, relying on quasitherapeutic relations and using all sorts of directives, in print he would have to rely increasingly on much more explicit and visible forms of authority because direct, face-to-face interaction was not available to the analytic community at large.

The emergent visibility of authority provides an important key to understanding the dynamics of the Freud–Stekel relationship. In face-to-face dialogue, Stekel's antagonistic counterposition proved to be a useful instrument in exercising authority, but in print it was much more difficult to use, as the *Gradiva* study revealed. Freud used *Gradiva* to transpose the face-to-face interaction of his Vienna society into a mediated quasidirect contact with the audience in order to set up a tutorial discourse. Freud must have realized, however, that the very antagonistic yet also instructive dialogue with Stekel could be transformed into a quasi-dialogical form only with great difficulty; this is the reason he sought refuge in literary technique, a piece of "imaginative writing," when he referred to Stekel as the "one among his followers" with whom he discussed this subject. Freud's problem here was not that this technique failed to provide enough realism to make it work. On the contrary, his problem was that any oppositional position in a written tutorial discourse calls the teacher's authority into question much more forcefully than in spoken language, where the audience is present and in constant contact with the teacher. This forced Freud either to reject his opponents' viewpoint in unambiguous language or otherwise push him back into the patient's position. He wavered between the two and finally chose the first option, but either way it meant the end of Stekel's serviceability to psychoanalysis: Freud's antagonistic relationship with Stekel became a burden, a counterproductive situation now that his wielding of authority had become increasingly visible. It is not a coincidence that he broke with Adler at the same time under much the same circumstances, and it was also precisely this transformation of visibility of authority that shocked Jung and prompted him to break away from the master:

> [Your] technique of treating your pupils like patients is a *blunder*. In that way you produce either slavish sons or imprudent puppies (Adler–Stekel and the whole insolent gang now throwing their weight about in Vienna). I am objective enough to see through your little trick (Jung to Freud, 18 Dec. 1912).

A consequence of these changes in his authority was that Freud slowly began to withdraw from his role as teacher–analyst in the Vienna society. He increasingly required fiercer critique of viewpoints not his own and approval of those that were; hence his public recognition of Stekel's dream symbolism and his simultaneous critique of *The Language of Dreams*, in which these interpretations were published. The transformation of visibility can also shed new light on the *Zentralblatt* affair, as the more visible authority becomes, the less tolerable open opposition will be, hence Freud's demand that Stekel unconditionally obey him. The end of their relationship was predicted earlier that year when Stekel directly questioned Freud's judgement in the masturbation debates. Freud then had few alternatives but to expel Stekel from the movement.

The Stekel episode represents a particular phase in the early development of psychoanalysis: a transformation in social organization which, because of its rapid development and growth, replaced direct, face-to-face dialogical interaction with written (printed), quasidialogical, mediated interaction. More importantly, this transformation marks a shift from implied to visible forms of authority. These changes left their traces in Freud's published works, his relationships with his followers, and in their published writings. A discursive analytic approach has laid these bare to investigation and interpretation. In the next chapter we explore some of these traces in more detail.

Chapter 3

MASTURBATION AND NEURASTHENIA
Freud and Stekel in debate on the harmful effects of Auto-Erotism[1]

Leendert Groenendijk

Introduction

As we briefly discussed the problem of masturbation in the previous chapter, it became clear that this issue drove a deep wedge between Freud and Stekel. It sealed their friendship, in a sense. Here we want to pick that issue up again and explore the differences between the two somewhat further, by focusing on the principle as well as on the personal aspects connected to this problem.

Masturbation, as Thomas Laqueur pointed out in his brilliant work on solitary sex (2003), was not problematized before the Enlightenment and became a *medical* problem only in the eighteenth and nineteenth centuries. In the early part of the twentieth century the "masturbator's delusion" was still very much a reality (cf. Kretschmer, 1966, pp. 80–95). Physicians were often confronted with patients believed to be suffering from the deleterious effects of masturbation (also labeled onanism, self-abuse, autoerotism, sexual self-stimulation, short-circuited love, and self-pollution, to name a few synonyms (cf. Gay, 1984, p. 295; Laqueur, 2003)). However, a dramatic change in the current attitude toward masturbation was soon to come. This was largely due to the work of sexologists such as Henry Havelock Ellis and

[1] This chapter is an adapted version of an article that first appeared in the *Journal of Psychology and Human Sexuality*, 9(1997), 71–94.

Albert Moll, whose research revealed no significant differences between healthy and mentally ill people with regard to autoerotic practices (Sulloway, 1979, p. 185). In the course of time, psychoanalysis was to join the attack on the oppressive attitude toward masturbation and effectively helped to free this sexual act from its atmosphere of secrecy and superstition, and turn it into discussible behaviour. An educational offensive was launched with books and leaflets and especially through the influential *Zeitschrift für psychoanalytische Pädagogik* (that ran from 1926 to 1938) in order to win adherents to a lenient attitude toward masturbation, which would ultimately prevent people from developing irrational anxiety and guilt (cf. Rehm, 1968).

This liberal stand in matters of autoerotism had, however, not always been prevalent among psychoanalysts. The originator of psychoanalysis did much to liberate sexual life from the constraints of Victorian moral opinion, but in the matter of masturbation Freud was far from revolutionary. With authors such as Krafft-Ebing, Eulenburg, and Rohleder, Freud emphasized the danger of masturbation because it was perceived as the cause of nervous pathology (Mortier et al., 1994, p. 840; see also Szasz, 1970).

It was Stekel who first pointed out the harmlessness of this practice. We surmise that the aforementioned liberal approach to masturbation among psychoanalytic pedagogues in the German-speaking world was largely the result of the arguments of Wilhelm Stekel. However, in those circles he was not acknowledged as the pioneering, liberating modernist *in sexualibus*. In the English-speaking world his progressive ideas on masturbation became mainly known through acts of "intellectual piracy," according to Emil Gutheil (in Stekel, 1950c, p. 3). The English translation of his provocative treatise on autoerotism was already completed in 1922, but publication was postponed until 1950. In the accompanying preface to that book, Gutheil urgently invited the reader to acknowledge Stekel's spiritual property and to honour him as the "Savior of Youth." We think that the failure to appreciate Stekel's merits must be related to the fact that members of the Freudian circle were expected not to mention the reckless man who had developed his views on the psychogenesis of the neuroses and the harmlessness of masturbation in overt confrontation with the teachings of his master (Mortier et al., 1994, p. 842, 847).

In this chapter we elaborate on the arguments behind the orthodox view on masturbation as defended by Freud and several of his disciples in the meetings of the Vienna Psychoanalytic Society, and present an exposition of Stekel's incessant taking the edge off that opinion, culminating in the latter's presentation in 1912, in which he boldly stated his ideas on the harmfulness of Freud's attitude toward masturbation.

Stekel's lecture was one of a series of rebellious, self-marginalizing acts. We argue here that the issue of masturbation played a major role in the alienation between Freud and Stekel. We attempt to explain why their conflict turned out to be unsurmountable: the thesis that we develop here is that in Freud's irrevocable rejection of Stekel and his particular point of view, a self-protective mechanism may have been at work (cf. Krüll, 1986).

Freud on the Sexual Etiology of Neurasthenia

When Freud presented his first psychological articles, one of the clinical pictures that often appeared in the consulting rooms was that of "neurasthenia." The first physician to write about this medical phenomenon was an American, George Miller Beard, who described it in 1869 in a magazine as an illness caused by "exhaustion of the nervous system" (Bromberg, 1963, p. 152; Jackson, 1986, p. 183). His views were adopted and adapted by renowned psychiatrists in France and German-speaking countries (Shorter, 1992, p. 221f).

In contrast to America, where modern civilization was held to be primarily responsible for the development of neurasthenia, in Europe hereditary constitution was believed to be the most important factor in the origin of a "nervous condition" (Drinka, 1984, p. 216, 218). Patients with a weak nervous constitution were thought to be predisposed to respond badly to modern hectic life. Treatment was in line with this view: in addition to medication, doctors prescribed diet, physiotherapy, electrotherapy, massage, bathing and, above all, rest (Gay, 1988).

"Neurasthenia" was soon treated as a label to be stuck onto all sorts of nervous complaints. It is to Freud's credit that he tried from the outset to arrive at a differential diagnosis, of neurotic symptoms, and that he tried to give explanations which were both comprehensive and simple. Freud launched the insight that sexuality played the main role in the etiology of all neuroses, both in actual neuroses as well as in psychoneuroses. The psychoneuroses, which included hysteria, phobia, and obsessional neurosis, were based on a defence against sexual representations from early childhood (cf. Moore & Fine, 1990, p. 154f). The actual neuroses had to do with a current, somatically induced, distortion of sexual life (Moore & Fine, 1990, p. 5). Thus, the concept of actual neurosis provided a physiological basis for symptoms.

In the early 1890s Freud differentiated between two types of actual neurosis: "neurasthenia proper" and "anxiety neurosis" (*Angstneurose*). Sufferers from "genuine" neurasthenic symptoms were tormented by fatigue, splitting headache, indigestion, tingling sensations, backaches, and weak potency. The usual symptoms of the victims of anxiety attacks were accelerated breathing, increased heart rate, palpitations, sweating, a fear of losing consciousness, and impending death. Freud contended that people with neurasthenic or anxiety symptoms hardly ever had a "normal" *vita sexualis* (heterosexual intercourse without contraceptive measures). People manifesting neurasthenic symptoms in the strict sense reached orgasm mostly through excessive self-gratification (masturbation), or suffered from nocturnal emissions, while sufferers from anxiety neurosis were predominantly practicing coitus interruptus or condomatosus. In this connection, it should be pointed out that Freud considered neurasthenia to be a man-made illness. The "hysteroneurasthenic" symptoms in women (who Freud believed were not normally "spoiled" by masturbation) were thought to be the result of inadequate sexual practices by the male (Krüll, 1986, p. 11f; cf. Showalter, 1987, p. 135ff).

Freud set himself up as a sharp critic of the current therapies for neurasthenia. "Kneipp-cures" and the like were rejected as useless. In contrast to the psychoneuroses, neurasthenic pathology was not amenable to psychoanalytic treatment. Neurasthenics were believed to be sufferers from sexual inadequacy (*Sexualitätskrüppel*). They had "poisoned" their own organism and had to be persuaded to give up this injurious practice. However, abandoning the masturbation habit would lead to anxiety neurosis (Freud, 1895a, p. 101f). The way out of this Catch 22 situation was to adopt "normal" methods of sexual discharge. The most important service which could be rendered to mankind was to prevent masturbation (and its damaging effects) by allowing youth greater (hetero) sexual freedom (Freud, 1898, p. 277f; see also Freud's famous attack on the oppressive role of "civilized" sexual morality, 1908a).

After 1920, neurasthenia as a medical diagnostic classification rapidly lost its popularity among physicians. The "medical paradigm" for interpreting nervous symptoms would become increasingly challenged by the "psychological paradigm" (Shorter, 1992, p. 232). And while Freud admitted that he hardly ever saw a neurasthenic, he would never give up or recant his ideas about the connection between neurasthenia and masturbation, but instead simply declared that he had lost interest in the actual neuroses. Thus the "Onan-myth" (Drinka) continued to dominate Freudian thinking about actual neuroses until well into the 1920s (Krüll, 1986, p. 14; see also Nagera, 1991). Still, the orthodox psychoanalytic viewpoint would change. What was Stekel's influence on this conversion, if any?

Stekel on the Role of the Psychic Conflict in the Etiology of Neurosis in the *Minutes*

In his *Autobiography* (Stekel, 1950a), and on several other occasions, Stekel stressed that it was thanks to his apologetic writing that the initial resistance to Freud and psychoanalysis had been overcome. It must be admitted that there is some ground for this claim. Yet it is more in line with what we know nowadays about Freud's reputation at the turn of the century (cf. Sulloway, 1979, 476ff) to describe Stekel's contribution as the further propagation of the Freudian gospel: through his popular writings the benefits of psychoanalysis were made accessible to a broad public (cf. his bibliography in Grinstein, 1958, 1882–1900).

But while Stekel accepted part of Freud's theoretical framework, he resisted most of the metapsychological speculations, yet was certain of having adopted the only true method for dealing with anxiety and other nervous disorders. As he began to visit Freud for therapeutic reasons, around the turn of the century, his convictions became embedded in a personal and beneficial acquaintance with the master, whom he came to admire as his "Christ."

Stekel was to make many more visits to Berggasse 19, mostly to obtain Freud's advice on certain patients. On one occasion, the enthusiastic disciple proposed that his master establish a discussion group in order to spread the light of psychoanalysis among doctors, literary men, and artists (Stekel, 1926, p. 542). This proposal met with Freud's approval, though he thought it wise to start with a small circle of familiar adherents (somewhat ironically labeled by Freud as "esoterics," (Stekel, 1926)) in order to prevent the spread of misconceptions. Once this circle was well-established, it would quickly be broadened to include others.

Since we have the *Minutes* (Nunberg & Federn, 1962–1975) at our disposal, we have access to a better understanding of the group dynamics of the Vienna society as a teaching institute (Bos, 1996). The *Minutes* are of vital importance for our insight into the early history of the introduction and spread of psychoanalysis. They provide us with an intimate picture of Freud as the leader of a movement. A leader who liked entering into conversation with his disciples, and who appreciated and stimulated the creative development and application of his ideas, but a leader who was also not willing to compromise on matters of principle. Thanks to these protocols we are also witness to the events and debates which made the departure of collaborators such as Adler and Stekel from the Vienna Psychoanalytic Society inevitable.

In this section we focus on the ninth protocol concerning the meeting of 5 December 1906 (*Minutes*, Vol. I, pp. 69–80). On that Wednesday evening a written text by Stekel on "Die Ursachen der Nervosität" (The Causes of Nervousness), was the subject of debate. Thirteen members were present. The discussion was opened by Eduard Hitschmann, who gave a critical summary of the essay (*Minutes*, Vol. I, p. 96f). The most important critical remarks, in addition to those of Hitschmann, were made by Freud and Reitler.

Before we discuss the reactions, we briefly outline the tenor of Stekel's essay which was published in the following year (Stekel, 1907a). Stekel wanted to make it clear that, in contradistinction to what was believed by many doctors, factors such as hereditary defect, mental stress, malnutrition, contagious disease and alcohol did not play a role in the development of neuroses. According to him, it was also incorrect to ascribe a main causal role to sexuality, as Freud had done. Stekel indicated that he, an otherwise "enthusiastic admirer of the great master," had been obliged to adopt a different course because of his clinical experience. All neuroses develop as the result of a "psychic conflict" between a repressed urge and human conscience. People become ill because their energy is being used for the unsuccessful repression of instincts. To illuminate his view, Stekel pointed to the problem of masturbation. How many people masturbated in their youth and nevertheless stayed healthy, he asked retorically. In order to cause a neurosis, a second, purely psychic factor has to join masturbation: the "psychic conflict."

> The auto-erotist or masturbator who finds himself in an ongoing struggle with sexuality, since he resolves to remain chaste and does not succeed, becomes nervous, as he spills an enormous quantity of energy in carrying out this inward struggle. His life is a continuous struggle between desire and anxiety (Stekel, 1907a [1933, p. 21]).

In the preface to the published essay, Stekel remarked that professor Freud would not give his assent to several passages in his pupil's book. He already saw "the great expert on inner life" shaking his head, his displeasure aroused. Thanks to the *Minutes* we know that Freud actually did vent his dissatisfaction with Stekel's pamphlet. He agreed with Hitschmann's critique, who had contended that the concept of psychic conflict does not offer a sufficient explanation for the genesis of neuroses (*Minutes*, Vol. I, p. 71f).

Reitler went further than all the other critics. He called Stekel's essay itself a "neurotic symptom": the author's life seemed to have been disturbed by a "wave of asexuality" (*Minutes*, Vol. I, p. 73). The full content of Reitler's comments is known because his contribution (added, typewritten, to the handwritten minutes) has been printed in full (*Minutes*, Vol. I, pp. 73–80). Thus Reitler points, not without humour, to a derailment of the otherwise competent stylist, who had written: "Freud, giving new life to an old superstition [sic], has ascribed to sexuality the main role in the etiology of the neurosis" (*Minutes*, Vol. I, 75). For the time being Reitler wanted to believe that this, indeed, was merely a stylistic lapse and not a "Fehlleistung": that the author did not want to maintain that Freud's discoveries were no more than old superstitions in a new, scientific guise. At the end of his paper, Reitler, by way of conclusion, repeated his earlier opinion that Stekel, who should have known as a Freudian that the neurotic's problems are ultimately related to improperly repressed sexual problems, but who swapped this deep insight for the opinion that conflicts lying at the surface are the cause of mental illness, had ostensibly written a text which could only be conceived of "as a neurotic symptom of defence against the author's own sexual instinct" (*Minutes*, Vol. I, 80).

This rather indiscreet and immaterial critique seems to have hit a sensitive spot in Stekel. It led him to make a public confession: he admitted having written the essay in a state of deep depression, "he had forced himself to write it, and he himself does not consider it good" (*Minutes*, Vol. I, p. 73). Notwithstanding this humble confession, Stekel doggedly upheld his view that masturbation as such did not cause physical damage resulting in neurasthenia; the injurious influence had to be ascribed to the psychic conflict only. This standpoint was iterated on every possible occasion. In the report of a discussion as early as 18 November 1908 we read about "the old [sic] controversy with Stekel about masturbation," and we are informed that "Freud's opinion still differs from Stekel's" (*Minutes*, Vol. II, p. 61).

The Discussions on the Harmfulness of Masturbation in the Vienna Psychoanalytic Society

The topic of "Onanie" had been touched upon in the Vienna Psychoanalytic Society several times. In 1907 a first attempt was made to deal with the problem in a discussion, introduced by Reitler, of Frank Wedekind's drama "Frühlings Erwachen" (*Minutes*, Vol. I, pp. 111–118). In 1909 Rank read a paper on the psychology of pathological lying, in which he sketched the analogy between anal and masturbatory character (*Minutes*, Vol. II, pp. 195–206). Apparently, the outcome of those discussions made clear that a new attempt had to be made. In the literature, reference is usually made to the "Onanie-Diskussion" of 1912, while the preceding discussion in 1910 seems to have been forgotten (e.g., Reich, 1951; Rehm, 1968, p. 109f).

On one evening in May and two evenings in June, 1910, the theme of the harmfulness of masturbation was discussed by several permanent members of the society and some guests. Of the permanent members present, Alfred Adler, Paul Federn, Sigmund Freud, Josef Friedjung, Carl Furtmüller, Max Graf, Margarethe Hilferding, Eduard Hitschmann, David Oppenheim, Otto Rank, Rudolf Reitler, Isidor Sadger, Maximilian Steiner, Wilhelm Stekel, and Victor Tausk took an active part in the discussion (see Mühlleitner & Reichmayr, 1992 for biographical details). The whole endeavour was intended to establish a more unified viewpoint on the issue. Everybody knew, of course, that the gravest and most important differences were to be found between Freud and Stekel. But in order to reach a more unified view, one of the main participants had to change or even withdraw his basic assumptions. Since this was not likely to happen, the question arises as to what the merits of such a discussion could have. What use could the contributions of the other members of the group have? Could they be used as a lever to force Stekel to give up his ideas, or would they have another discursive function?

We start with the meeting of 25 May 1910, the first session in the "harmfulness of masturbation" discussions (*Minutes*, Vol. II, pp. 542–552). After pointing to the remarkable fact that many physicians now stressed the harmlessness of masturbation, while patients were convinced of its dangerous effects, Freud formulated the main questions to be answered by the members of the group: why does masturbation seem to be injurious in some cases and not in others? And is masturbation as such harmful, or something at the bottom of it? (*Minutes*, Vol. II, p. 542).

The first speaker was Hitschmann, who at once showed himself to be a firm defender of Freud's position. According to him, the actual neurosis of neurasthenia is a consequence of increased masturbation. Doubt about the harmfulness of masturbation is partly a resistance to Freud's teachings on the sexual origin of neuroses (*Minutes*, Vol. II, p. 543). Everybody will have thought of Stekel. However, he did not counterattack Hitschmann and other critical discussion partners directly. Instead, he read some passages from a book (published in 1873) by the Russian physician von Gutceitt, according to Stekel a "precursor" of Freud, who (evidently in contrast with his "follower") regarded masturbation as harmless and who was opposed to preventing this means of sexual gratification (see also Stekel, 1910, p. 68f). He then stressed that most of the speakers had forgotten to deal with the main problem. According to Stekel, the whole issue had been raised by Freud because he wanted to reexamine the nature of neurasthenia: whether it was caused by masturbation or not (*Minutes*, Vol. II, p. 551f).

Freud himself did not confirm Stekel's formulation of the problem (namely that it was not neurasthenia but masturbation that had to be reexamined) but instead admonished the members not to ignore the usefulness of masturbation, in addition to considering its harmfulness (*Minutes*, Vol. II, p. 552).

On 1 June 1910 (*Minutes*, Vol. II, pp. 554–565) the continuation of the discussion on the harmfulness of masturbation was opened by Sadger, who acted as chairman. He reminded those present of Freud's remark not to ignore the usefulness of masturbation and its practice by women. It was Stekel's turn to give an exposition of his view. From the minutes we can gather that Stekel took the opportunity to advocate his view that neurasthenia does not exist and that what is involved with masturbation is a solely psychic conflict (*Minutes*, Vol. II, pp. 554–556). We discuss his view in detail in a separate section, and limit ourselves here to an impression of some of the reactions.

Most of the members did not approve of Stekel's exposition. According to Steiner "what was good was not new, and what was new was not good" (*Minutes*, Vol. II, p. 557). Only Adler agreed openly with Stekel, albeit "as far as the nonexistence of neurasthenia is concerned" (*Minutes*, Vol. II, p. 557). Adler, the supporter, soon turned into a competitor however, as he and Stekel became involved in a quarrel about the discovery of the conception of the unity of the neuroses. Adler rejected Stekel's claim to priority by reminding him of the fact that in his 1908 book *Nervöse Angstzustände und ihre Behandlung* (Nervous Anxiety States and their Treatment) he had made a distinction between anxiety hysteria and anxiety neurosis (*Minutes*, Vol. II, p. 560). It is to be suspected that Freud and his orthodox followers would have been amused by this painful competition between the two dissidents.

In his final remarks, Freud not only summarized the dangers of masturbation, but also its advantages (*Minutes*, Vol. II, pp. 561–563). He mentioned, among other things, the reduction of sexual tension that made it possible to devote oneself to other tasks.

Stekel was eager to declare that he was pleased that Freud had accepted some of his thoughts. However, his joy did not last for long. Federn asked Freud to set out his viewpoint with regard to neurasthenia. Freud consented; his remarks (which are discussed in a subsequent section as part of a summary of his published reaction to Stekel's view) did not corroborate Stekel's conclusion. He declared that he had changed his views on neurasthenia much less than was believed and that he was still convinced of the difference between actual neuroses and psychoneuroses (*Minutes*, Vol. II, p. 564).

The final discussions on masturbation took place on June 8 (*Minutes*, Vol. II, pp. 566–572). Adler covered the topic from his particular point of view (*Minutes*, Vol. II, p. 567f). The harm that appeared in masturbation, the feeling of guilt etc., could be traced back to before this habit had begun. Adler thought he could prove his insight, but the adherents to Freud's view refused to be convinced. Federn declared that he still shared Freud's opinion and not Adler's, "even though certain observations would seem to speak against it" (*Minutes*, Vol. II, p. 569). As was to be expected, Adler contended that masturbation was just one form of "the masculine protest." Krauss, editor of *Anthropophyteia* (a journal devoted to the history and ethnology of sexual morality), and present as a guest, raised his doubts. He pointed to masturbation for magical purposes and as a children's game in Bosnia (*Minutes*, Vol. II, p. 568f). Even in those cases, however, Adler was able to find masculine protest. Stekel objected that Adler had again fallen into the error of applying his principle to everything. Some of the other speakers who did not even fully agree with Adler nevertheless reproached Stekel for this criticism.

At the end of the evening, Freud's desideratum, masturbation in women, was finally treated, albeit only by some "indefinite observations" from Margarethe Hilferding (*Minutes*, Vol. II, p. 571f). Thus this first round of discussion on the problem of masturbation petered out in a somewhat dissatisfying vein: the confrontation between the opposing viewpoints had not resulted in a clear cut either- or viewpoint. Freud decided that it would be better not to publish the discussions and to continue the study of the subject. "Perhaps," wrote Laqueur (2003, p. 367), "Freud was trying again to avoid an answer."

It took a year and a half before the society was ready to approach the subject again. On three Wednesday evenings in 1911 and five in 1912, the members of the Vienna Psychoanalytic Society again debated the issue of masturbation. Fourteen papers were presented by (in order of appearance): Isidor Sadger, Eduard Hitschmann, Victor Tausk, Otto Rank, Josef Friedjung, Sándor Ferenczi, Rudolf Reitler, Wilhelm Stekel, Maxim Steiner, Paul Federn, Hanns Sachs, Gaston Rosenstein, Bernhard Dattner, and Sabina Spielrein, the only female contributor to the discussion. On 24 April 1912 Freud dealt with the results in a retrospective (*Minutes*, Vol. IV, pp. 92–96). The record of those discussion evenings is included in the third and fourth volumes of the *Minutes*.

Stekel delivered his paper on 7 February 1912 (*Minutes*, Vol. IV, p. 36f). Seven lecturers had already preceded him. The majority more or less disagreed with Stekel, who in turn did not appreciate their approach i.e., one of masturbation as an illness.

During the first discussion evening, when the debate had readily moved to pathological masturbation and one of the guests had stimulated the imagination by pointing to a particular bathtub, Hitschmann piously warned that one should not forget to deal with the relationship between masturbation and neurasthenia. A short-circuit to Stekel, so to speak, because from his introductory lecture, Hitschmann himself proved not to have forgotten this relationship in the least, as he attacked Stekel's point of view broadside (*Minutes*, Vol. III, pp. 336–338). Rank recorded Stekel's reaction to this attack as follows: "Stekel agrees with almost nothing and, instead, condemns Hitschmann's reactionary viewpoint" (*Minutes*, Vol. III, p. 338).

A promising kick-off of the discussions. And Stekel continued his critique when he said that lecturer Tausk had bitten off more than he could chew and had been talking in psychoanalytic platitudes. Tausk in his paper (*Minutes*, Vol. III, pp. 339–343) had pointed to the relationship between one's choice of neurosis (*Neurosenwahl*) and one's favorite masturbation zone. It is notable that Freud wished to protect Tausk against critique, at least in so far "as many of us also think that the central problem in the choice of neurosis does, in fact, lie in the question of masturbation" (*Minutes*, Vol. III, p. 346). A sharpening of the swords.

Friedjung's paper (*Minutes*, Vol. IV, p. 20f), which was based mostly on clinical data, was highly appreciated by Stekel. Nevertheless, according to Stekel, his observations were incomplete. Friedjung had concluded erroneously that children began to masturbate at four months old, whereas Stekel had demonstrated in his first publication, *Über Coitus im Kindesalter* (On Coitus in Childhood; Stekel, 1895), that masturbation, genital and anal, could be observed during the very earliest period of life (*Minutes*, Vol. IV, p. 22).

Stekel's own presentation, to which we turn in a moment, had been preceded by that of his long standing adversary Reitler. Even before starting his paper, Reitler said that he felt obliged to criticise Stekel's opinion that neurasthenia is a psychogenic illness: "not everything that can be influenced by psychological means must for that reason be psychogenic" (*Minutes*, Vol. IV, p. 35).

With respect to the lecturers who followed, it is of interest to note that, of the four, only Sabina Spielrein's lecture could win Stekel's approval. According to the protocols, his response to the last but one, in which the vicissitudes of masturbation were traced from earliest childhood to the years of maturity, could be summarized as follows: "Stekel considers Dattner's statements to be wrong; in most cases a period of latency does not exist. If children become (damaged), it is not as a consequence of masturbation, but rather of their giving up masturbation" (cf. *Minutes*, Vol. IV, p. 78).[2]

This time, the divergent views fully clashed, for all to see. This time also, Freud had decided to salvage something from all his group's talk (Laqueur, 2003, p. 367): he now thought the time was ripe to publish the debate.

Stekel's Printed Contribution on Masturbation and Freud's Critical Retrospective

Therefore, in 1912, not long after the discussions had ended, the second volume of the *Diskussionen des Wiener psychoanalytischen Vereins* was published (Dattner et al., 1965), containing fourteen contributions to the debate on masturbation. Almost all of the main contributors of 1911/1912 are present in this volume, including Stekel, who had delivered his paper to the society on 7 February 1912. He was represented in the discussion volume with a lengthy paper in which he presented his views unaltered and undiluted. In this section we summarize Stekel's definitive treatment of his position, which evoked much resistance and was defended with at least as much vigour (Dattner et al., 1965, pp. 29–45; also in Lütkehaus, 1992, pp. 246–261). We then go on to discuss the main points of Freud's published retrospective of the debate on the harmfulness of masturbation.

Stekel opens his paper by stating that according to him, Havelock Ellis's expression of "autoerotism" is preferable to the antiquated and maligned term "onanism." His definition of autoerotism is as follows: every sexual act that is carried out without the cooperation of another person. In other words: autoerotism is an unsocial activity. Essentially, it is a manifestation of narcissism because the pleasure is derived from one's own body.

Everybody has masturbated, said Stekel. This thesis can be maintained once we understand that there are sundry varieties of masked masturbation, such as nose-picking and other infantile forms of pleasure. Given the universality of

[2] English translation amended by the authors; compare the German edition of the *Minutes* in Nunberg & Federn (1976–1981).

masturbation, one may infer its harmless nature and, consequently, the impossibility of its being the cause of anxiety neurosis. Yet one often observes a range of harmful symptoms after masturbation. Masturbators complain about headaches and their being tired out, etc.: a condition that Ferenczi had labeled "one-day neurasthenia" (*Eintagsneurasthenie*). According to Ferenczi, the temporary neurasthenic symptoms were a physiological effect of masturbation. Stekel wanted to oppose this medical view, and maintained his conflict-psychological explanation. The onanist did overcome his religious, moral, and hygienic inhibitions, but after the act they made themselves known as reproaches leading to depressiveness.

Stekel confirmed Freud's observation that (anxiety-) neurosis breaks out after abstinence from masturbation. However, he objected to the view that there is a causal relationship between masturbation and neurosis. Neurosis is, in fact, the *consequence* of abstinence, not the other way around. As an illustration, Stekel referred to the case of a thirty year old doctor's wife. For her, masturbation was the only enjoyable form of sexuality. She had been practicing it without any problems from her fourth year onwards. After reading anxiety-provoking literature, however, she became terrified and on her husband's advice stopped the habit. Then a heavy depression ensued which brought her almost to the point of suicide. Some years before, in 1910, when discussing the subject of suicide, this case had also been reported by Stekel. At that time he had explained the urge of this woman to kill herself as the ultimate act of autoerotism (in German, the expression for suicide bears an ambiguous connotation: *die Hand an sich zu legen*; see Adler et al., 1910, p. 37).

Stekel then moved on to excessive masturbation (labeled *Onanismus*). His remarks on this topic were rather provocative. According to him, the excess of this sexual habit is not at all as dangerous as it was claimed to be. Where in earlier circumstances Stekel would have confined himself to be declaring that he made no bones about moderate masturbation, he now sketched, with barely concealed admiration, the sperm-rich days and nights of excessive masturbators. He preferred to look upon them as sexual athletes rather than as patients. According to Stekel, who apparently admired Don Juan even more than Onan, such people have a very strong sexual constitution and it is best for them if they follow their nature, preferably with a lover who knows how to esteem their potency.

However, "normal" sexual intercourse is not granted to everybody, and here Stekel mentioned homosexuals. For those who dare not live their life to the full, masturbation and homosexual fantasies are an indispensable and adequate substitute. In cases such as these one should not urge the patient to seek a (hetero)sexual partner. In sexual matters, therapists should not behave like defenders of "normalcy" because such a canon does not exist in reality.

Autoerotism has to be valued highly because of the fantasies which accompany it enable people to play all sorts of roles. This asocial activity has an important social function as it protects society against crimes against morality being committed. The masturbator stages his sexual crimes only in his fantasy. Suppression of this habit would increase the number of sexual crimes very rapidly.

Before leaving the topic of masturbatory fantasies, Stekel mentioned the fact that they frequently reveal that masturbation is a regressive phenomenon in the Freudian sense. People masturbate for instance with the fantasy of the penis as the nipple that is being milked. In this connection Stekel pointed to the fact that men often speak of "milking" (*melken*) to indicate the masturbatory act. So the masturbator relives the primary and strongest (oral) lust of man: the pleasure of sucking. This might explain, said Stekel, why it is sometimes impossible to wean people from the masturbation habit.

Stekel dealt extensively with the complex relationship between masturbation and feelings of guilt. Only when guilt becomes attached to masturbation will neurotic symptoms, which had been described as the consequences of masturbation, arise. Experience had taught him that information about the harmless character of moderate masturbation and the analysis of fantasies accompanying this indulgence did not always bring the expected relief. What was the case? Masturbation had taken over other—unconscious—reproaches which were much more painful than the reproaches against masturbation itself. A shift of guilt had taken place. Masturbation had become the reservoir of all varieties of guilt. As a therapist, one always has to make sure whether the masturbator's complaints might have a deeper origin and should be dealt with there.

That the masturbator is not easily delivered from his guilty conscience has still another reason. Stekel cited Otto Rank, who was the first to point out (in his book *Der Künstler*) that man has the tendency to increase his sense of gratification through the development of inner (artificial) resistances. Man is a fighter; since contemporary culture allows no outward fighting, we have to resort to the staging of inner conflicts. Because masturbation is forbidden (it is called a "sin" by the Church, because the act does not serve the holy duty of procreation), overcoming it has a high pleasure-value. The sense of guilt works as a stimulus and at the same time the stimulus operates as an inhibition (*Die Hemmung wird zum Reiz und der Reiz zur Hemmung*). So, according to Stekel, the principle of "bipolarity" in all psychic phenomena turns out to operate in the mind of the masturbator (Dattner et al., 1965, p. 41).

With Stekel's viewpoints now unambiguously brought into the open, it was up to Freud to respond. He did so in the final contribution in the collection of papers on masturbation. It is a retrospective, adapted from his concluding comments on the discussions of 1910 and 1911/1912 (in Dattner et al.,1965, pp. 132–140; Freud's contribution without the others in Freud, 1912b, pp. 243–254). According to Freud, the discussions led to a more unified view on masturbation. There was full agreement on the meaning of masturbation fantasies and of the role of guilt. Yet it could not be denied that some differences of opinion also remained, especially with regard to the harmfulness of masturbation and its explanation. The merit of Stekel's critical remarks, wrote Freud, was that other participants were stimulated to present their viewpoints clearly.

After having briefly remarked that most writers neglected to treat masturbation in connection with respective developmental periods, and after a positive comment on Reitler's contribution, Freud quickly came to business: three quarters of his contribution is devoted to a discussion of Stekel's position. Freud almost immediately concludes that he found himself in the majority against Stekel with regard to the relationship between masturbation and neurasthenia. Stekel carries the psychogenic point of view too far, wrote Freud; he is wrong in denying the toxic effects of masturbation and, indeed, the existence of neurasthenia. Freud maintained his conviction that the neurasthenic does not belong on the couch. At least not to begin with, for Freud now left the door open to the possibility that psychoanalysis may be helpful in the treatment of persistent neurasthenic symptoms.

Freud also deplored the fact that the discussions had been narrowed down to the question of the injuriousness of masturbation, though he admitted that with this line of approach the public interest was being satisfied. Again, he cannot agree with Stekel, for whom the harmfulness of masturbation is virtually a "senseless prejudice." In addition to organic damage and weakening of potency, there are indeed harmful psychic effects. The masturbator forsakes reality, hindering the development of strong character traits. Masturbatory fantasies, especially, preclude the formation of healthy sublimations. The masturbator lives in accordance with the principle of lust and is thereby caught up in psychic infantilism, by which a neurotic development is being prepared. Otherwise Freud is ready to admit, with Stekel, that a positive aspect of the masturbator's fantasy is that grave perverse cravings are rendered harmless by it. Freud also discerns a positive aspect to the diminution of potency: a certain loss of it can mean a gain for society. He observes: "(...) masturbation, like so many other things, has 'les défauts de ses vertus' and on the other hand 'les vertus de ses défauts'" (Freud, 1912b, p. 252); a conclusion with which he seemed to mock Stekel's aforementioned principle of bipolarity. We return to this ironic remark in a later chapter.

The Onanism Issue: The Key to Understanding Freud's Dissociation from Stekel

So the discussions were slowly directed toward the conclusion that Stekel had made himself impossible, and Freud used the onanism dispute to make that clear once and for all. The debate provided incontrovertible proof of Stekel's unorthodoxy, of his disloyalty and his irresistible urge to make his own position known. It was, however, also a painful demonstration of the isolated position Stekel had come to occupy within the Vienna Psychoanalytic Society, especially after the departure of the Adlerians. The publication of the papers on masturbation was intended, according to the preface, to give the readers an indication of the open-minded climate within the Vienna Psychoanalytic Society. That must be considered a particularly ironic statement, since this document could just as well serve as a public testimony of Stekel's separatism and of the accuracy of the conclusions about the differences drawn by Freud.

With friends like this, who needs an enemy? With evidence like this, who needs a pretext to make Stekel leave the Society? Between publication of these debates and the official cause of his break with Stekel lay but a few months. In November of that year Freud claimed that Stekel had acted unethically as chief editor of the official organ of the *Internationale Psychoanalytische Vereinigung*, the *Zentralblatt für Psychoanalyse*. The minutes of the meeting of 6 November 1912 carry the entry: *Stekel ausgetreten* ("Stekel resigned;" an entry that does not appear in the English edition).

The masturbation issue was not blown up by Freud for instrumental reasons only, as the fundamentals of his structure had been attacked directly. It was intolerable that precisely Stekel was responsible for this. Stekel himself was convinced that this was the key to explaining how he had got himself so seriously into Freud's disfavour: "Freud could not forgive me for pointing out the harmlessness of masturbation, and for having proven his conclusion that neurasthenia, brought about by masturbation, pointing out the harmlessness of masturbation, and for having proven wrong his conclusion that neurasthenia is brought about by masturbation" (Stekel, 1926, p. 549).

But why, though, was Stekel's refutation so particularly intolerable? In order to understand this, it will be necessary to go back to around 1900, when Stekel underwent a short analysis with Freud. Freud remembered Stekel as a patient who had confessed suffering from potency difficulties (cf. Stekel, 1950a, p. 107, 123). In a covert autobiographical case history in his book on male impotence, there is a hint that the particular nature of his own sexual difficulties (for which he eventually sought Freud's help) may have been ejaculatio praecox (see Stekel, 1971, Vol. I, p. 126; see also the chapter on Stekel's *Lehrjahre* in this volume, for a fuller discussion of this autobiographical case history).

We tentatively reconstruct the Freud–Stekel therapeutic dialogue as follows: During his brief analysis, Stekel told Freud that he had already suffered earlier from weak potency and also from morbid thoughts and anxieties about his masturbation and nocturnal emissions (cf. Stekel, 1971, pp. 116–127; Stekel, 1929, p. 23f). Freud may have recognized in himself some of his own problems and Stekel may have guessed this. Freud advised his patient/colleague-to-be to give up his masturbation habits, and it worked. It relieved Stekel of his sexual problems. But not because he followed this advice, but precisely because he *didn't*. He understood that he and Freud were bound by a shared secret *in sexualibus*. He had gained an insight into his inner inhibitions, his feelings of guilt and anxieties, into the psychogenic cause of his (potency) problems, because he related them to Freud's inner inhibitions, anxieties, and feelings, etc., and concluded that all he had to do was the exact opposite of what Freud did. And so he became a defender of the harmlessness of masturbation thanks to the "redressement psychique" brought about by psychoanalytic therapy!

This reconstruction, speculative as it is, may not correspond at all to how Freud would have thought about it, nor, indeed, Stekel himself. The reason we present it is to make perceptible at a psychological level what can be found at a discursive or interactive level, namely a dialogue wound up in antagonism and opposition. For Stekel would afterward (after his break with Freud) always insist on his master's faults, the "blind spots" in Freud's thinking, while Freud would develop such an abhorrence of Stekel that it made Wittels wonder why he had put up with him for so long.

Indeed, whatever it was that Stekel had revealed during his therapy, it was important to Freud. Stekel's revelations were a welcome and significant clinical confirmation of Freud's theory about the pathological effects of masturbation, which was based on the interpretation of his own experiences with the practice (Buelens, 1971, p. 228f). Freud was unwilling to accept Stekel's later remarks about the harmlessness of masturbation because those later remarks implied that the relationship between masturbation and neurasthenia and between masturbation and impotency had no "empirical" foundation, but would exist only in Freud's imagination. He tried to make Stekel's scientific unmasking of the "Onan-myth" harmless by transforming it into an act of a person who lacked integrity. "One day when I am no more—my discretion will also go with me to the grave—it will become manifest that Stekel's assertion about the harmlessness of unrestrained masturbation is based on a lie," Freud wrote in a letter to Fritz Wittels (15 Aug. 1924, in *The Letters of Sigmund Freud*, 1960). Stekel was a betrayer, who had violated a secret bond.

Researchers (e.g., Oerlemans, 1949, p. 41ff) have wondered why Freud, who spent his life in constant elaborations and refinements of his concepts, was unable to free himself from his own theory and would stick to his viewpoint to the bitter end. Well, he was too much of an old-fashioned medical doctor, wrote Laqueur, to let go of the medical somatic model, and besides, deep down, he never overcame his bourgeois adolescence anxieties with regard to masturbation (Laqueur, 2003, p. 381, 387). That may be true, but it does not answer why he could never give up his views on masturbation—and succeeded in moving beyond other, equally sensitive issues.

To answer this question we return to his discussion with Stekel, which aroused in him such a considerable aversion. What follows is a speculative conjecture, partly based on Marianne Krüll's interesting hypothesis about Stekel's unmasking of the self-protective function of Freud's theory of the actual neuroses (Krüll, 1986, p. 190).

Following Freud's train of thought, we may say that if Stekel's approach was correct, "neurasthenia" did not exist. But that could not be true, for if it were, Freud, who described himself as a sufferer from neurasthenic and anxiety neurotic symptoms (cf. Buelens, 1971, p. 225ff; Krüll, 1986, p. 17, 19; Shorter, 1992, p. 222), would have to reexamine the biological explanation of his own complaints, even reexplore his own psychological complexes. And that may very well be what Stekel told him while he was in analysis with him. It certainly was what he wrote, repeatedly, afterward: that the master needed analysis himself. Indeed, Freud had never been analyzed by anyone. Would he not stretch out on the couch, then and be analyzed ... by Stekel?

This annoying attempt by Stekel to equal or even outdo the master certainly did his relationship with Freud no good. In it we find the seeds of their priority struggles which would surface in the 1910s and lead Freud to eliminate Stekel from his circle. The elimination of Stekel came at a moment when psychoanalytic discourse had evolved from an open, underground "tongue" to a formalized language, that no longer needed opponents to articulate its identity, but rather wished to silence its inner doubts. But it also came at a moment when Freud's marital sexual activity had diminished while his erotic fantasies still continued for years (cf. Gay, 1988, p. 163; Krüll, 1986, p. 18f) and he therefore was tempted to commit "Onan's sin." Perhaps his rebuff of Stekel's enlightened views also represented a taboo instilled in his childhood by his father, Jacob Freud, who had discovered his son's "self-abuse" and who threatened him with castration if he ever played with his genitals (Krüll, 1986, p. 110ff, 142; cf. Oerlemans, 1949, p. 44). If so, the attempt to silence Stekel also stood for the wish to terminate an inner conflict of the "neurasthenic" father of psychoanalysis as well. It is interesting to note that he could not banish Stekel entirely: Freud abstained from reading Stekel's writings, but secretly commissioned his disciples to read and examine them for him (Freeman, 1968, p. 268; Roazen, 1992, p. 220). So the debate with Stekel never ended in Freud's mind, and he never missed an opportunity to point out verbally or in writing Stekel's sins, whom he came to think of as a "dirty person."

Chapter 4

CURING SOCIETY BY BETTER EDUCATION

Wilhelm Stekel on Parental Influences in the Causation and Prevention of Compulsion Neuroses[1]

Leendert Groenendijk and Johan Sturm

Introduction

Rapoport's *The Boy Who Couldn't Stop Washing* (1989) is not a fairy tale. Her book deals with one of the most incredible illnesses of the soul, with OCD, the obsessive-compulsive disorder, a condition marked by obtrusive ideas and compelling acts, which the victim, despite all his efforts, is unable to dispel. For example: six hours of scrubbing a day and still doubting whether one is clean (see Rapoport, 1989). In the older literature this bizarre disorder was subsumed under the heading of *neuroses*. Patients with obsessions and compulsions were assumed to have a premorbid ("anankastic") personality that is causally connected to the genesis of the disorder (see for instance Schneider, 1940, pp. 63–73). According to modern insights, however, there is no clear-cut direct link between the obsessive-compulsive personality and OCD, although it is granted that a small proportion of OCD patients also show traits of the extreme variant of this

[1] This chapter is a slightly adapted version of an essay that first appeared in *Paedagogica Historica*, *36*(2000), 653–673.

character style, the obsessive-compulsive personality disorder (OCPD) (Johnson, 1994, p. 269).

Obsessive-compulsive phenomena have been the subject of much debate and confusion in the scientific literature from the nineteenth century onward (Berrios, 1995). Then, for the first time, psychiatrists undertook to construct an empirical alternative to the old religious or rather diabolic attribution of the phenomena. There was much speculation on which organic factors might cause the symptoms to originate. Brain disturbances caused by fevers and hereditary-degenerative processes were thought to cause the disease. It was linked with the then popular concept of "neurasthenia" (exhaustion of the nervous system); obsessions or "fixed ideas" were also understood as manifestations of a condition called "psychasthenia" (mental exhaustion) by the French psychiatrist Pierre Janet, who related this condition to the *caractère scrupulo-inquiet* (Mayo, 1951).

Janet made an extensive study of 236 obsessionals in Paris, however, his meticulous research remained relatively unknown outside the French-speaking world. Most investigators were to follow Sigmund Freud, who built up a psychosexual theory on nervous disorders, including what he called the *Zwangsneurose*. Freud, who personally had some experience with obsessions, developed his thoughts on the subject over a period of thirty years.[2] Two of his most celebrated cases, those of "The Wolf-man" and "The Rat-man," suffered from severe obsessive-compulsive complaints.

Already in his earlier work, Freud demonstrated an interest in which preconditions were required in order to create a *Zwangsneurose*. In 1896 he assumed that a repressed traumatic sexual experience during childhood was the basis of a (psycho-)neurosis. Children who underwent certain premature sexual experiences passively were prone to develop a hysterical neurosis, while children who behaved actively would develop an obsessive-compulsive neurosis. Because many traumatic events turned out to be products of fantasy, Freud saw himself obliged to reject his so-called "seduction theory" and replace it by the Oedipus-complex theory as being the all-inclusive explanation of psychoneurotic pathology (see Israëls & Schatzman, 1993; Albach, 1993, for a discussion of this move).

In Freud's later and more well-known perspective, the basic problem of the obsessive-compulsive neurosis (OCN) was the ego's conflict with primitive (id) impulses and the commands of the superego (see for a discussion of Freudian views on the origin of OCN Fenichel, 1946, pp. 268–310, still further

[2] For his theoretical contributions to the theme, see especially Volume III of *The Standard Edition of the Complete Psychological Works of Sigmund Freud* (London: The Hogarth Press).

condensed by Robert Cawley in: Beech, 1974, p. 270f). A compulsive symptom may represent a condensation of primitive wishes and defences against them. The concept of pathological regression (of the libido) forms the cornerstone in the construction of OCN. After the ego has developed to a stage where drives are at a genital level, an intolerable increase in "castration anxiety," related to the so-called Oedipal situation, leads to a regression (of the libido) to the pregenital anal-sadistic level. The anal sadism combines with the Oedipus hostility felt for the parent of the same sex, and this leads to increased ambivalence of object relationships. The superego has an important influence at this stage, evoking anxiety and guilt which cause further conflict within the ego, hence so-called defence mechanisms (*Abwehrmechanismen*), e.g., affect-isolation and undoing, are called into play, which are, however, rarely sufficient in preventing the patient from an aggravating illness. According to Freud there was no route to OCN other than through the development (in connection with harsh toilet-training) of what he called the "anal-erotic character" (Rachman & Hodgson, 1980, p. 55).

Freud's publications stimulated many researchers and therapists to adopt an understanding of compulsive phenomena along "psychodynamic" lines, Wilhelm Stekel being one of the first doctors who had been converted to this approach. In this essay we want to analyze the ideas of this pioneer psychoanalyst on the nature, genesis, and prevention of what he termed the *Zwangskrankheit* or *Zwangsparapathie*.

Stekel's approach to the condition of OCN developed over a period of time and ultimately culminated into a *pedagogical* approach; his final view forms a "grandiose" attempt at redefining OCN as a social and moral problem—a task for educators and all those concerned with human welfare. In this respect, his work offers, as will become apparent, an illuminating insight into the psychoanalytic enterprise at making parents believe that their culturally induced (sexual) behavior is of crucial importance for both the explanation and the prevention of mental illnesses in the life of their offspring as children as well as grown-ups. In order to prepare parents for a more wholesome behavior, Stekel called on doctors and psychotherapists to become educators of parents. With his publications he contributed to the *psychoanalyzing* of the ancient art of child raising, which was going to have an enormous response among parents in several countries of the Western world during the greater part of the twentieth century (see Gross, 1979, pp. 247–276; Groenendijk & Bakker, 2002).

Stekelian Publications on OCN

Stekel, as well as his followers, have published several studies on the subject of OCN. In 1909, the same year in which Freud published his "Notes upon a case of obsessional neurosis," Stekel presented a paper to the *Vereinigung*

Wiener Mediziner (Vienna Medical Society) on "the psychic roots and therapy of compulsive states" (*Zwangszustände, ihre psychischen Wurzeln und ihre Heilung*). The paper was published the following year. It was an enthousiastic confirmation of Freud's view on compulsive states as being psychically motivated and meaningful (*sinnreich*). In 1912 Stekel published an article on the phenomenon of doubt. In the papers just mentioned we are introduced to several of Stekel's patients; his acquaintance with and interest in obsessive-compulsives is also visible in the cases of *Die Sprache des Traumes* (The Language of Dreams), the book on the interpretation of dream symbols he published in 1911. The last two volumes of the series on "the disorders of the instincts and the emotions," *Zwang und Zweifel* (Compulsion and Doubt), which appeared in 1927 and 1928, respectively, are devoted to the compulsive disease or *Zwangsparapathie* as Stekel preferred to name it (Stekel, 1927–1928).[3] The two compulsion and doubt-volumes are crammed with Stekel's patients, success stories as well as failures. From his case histories and interpretations it becomes clear that Stekel did not confine himself to what is nowadays labelled as OCD—there is an overlap with the condition which is currently called OCPD, the obsessive-compulsive personality disorder.

In 1930, in Baden–Baden in Germany, a medical congress for psychotherapists was held on the subject of OCN. The famous psychiatrist Ernst Kretschmer asked Stekel to deliver an introductory lecture. Among the other participants who presented a paper were Karen Horney, Ernst Simmel, Erwin Straus, Franz Alexander, and several of Stekel's pupils. In his lecture, Stekel gave an account of the pathogenesis of OCN as a socioculturally induced and pedagogically transmitted disease. Stekel's contribution, *Die Psychologie der Zwangskrankheit*, is included in the proceedings of the conference; in 1931 it was published in a German-speaking as well as in an English-speaking scientific journal. The paper is also included in his last book, *Die Technik der analytischen Psychotherapie*, published in 1938, of which an English translation entitled *Technique of Analytical Psychotherapy* was published in 1939 in England and the following year in the USA, and a revised translation, with an introduction by his pupil Samuel Lowy, in 1950 (see Stekel, 1950b, pp. 280–318, Chapter XX: "Psychoanalysis of Obsessional Disease").

Stekel's pupil Ernst Bien wrote an interesting bibliographic essay for Stekel's *Zwang und Zweifel* (Bien, 1927–1928), and had also presented a paper at the Baden–Baden conference, proving himself to be the right mouthpiece of Stekel's views in his lucid contribution to *Der Seelenarzt* (The Soul Doctor), the handbook on nervous problems of the "Institute for Active

[3] The older studies on compulsive states and doubt mentioned in the text, have been incorporated into Vol. I of *Zwang und Zweifel*. Stekel preferred the term "parapathy" to "neurosis," because not the nervous system, but the emotional life is afflicted.

Psychoanalysis" in Vienna (Bien, 1933). Gutheil, the best known follower of Stekel in the USA, gave an outline of his master's views on OCN in the preface of *Compulsion and Doubt* (1949), his (abridged) translation of Stekel's *Zwang und Zweifel*.[4]

OCN: A Social Disease on the Increase

Among historians of psychiatry it is a truism that both the clinical picture of mental illnesses and their incidence have been subject to change in the course of time (see Shorter, 1992). This insight is not a prerogative of modern historians, as can be illustrated by Wilhelm Stekel's fascination for the changing reality of neuroses. Stekel (1950b, p. 282) was convinced "that the parapathies (neuroses) are modified by social conditions." This was not a unique or extreme standpoint in those days; among his contemporaries were even advocates of a purely sociogenetic perspective on neuroses (such as Hellpach), as can be learned from Fischer-Homberger's well-known study of the history of the "traumatic neurosis" (Fischer-Homberger, 1975, p. 182ff).

In his student days and in the years of his general practice at Vienna most of the patients seen by Stekel were hysterics. Thirty years later hysterics were rarely encountered. The so-called hysterical attack in three phases as described by Charcot had almost vanished. On the other hand, there had been a rapid increase in sufferers from obsessive-compulsive complaints: "I see hundreds of cases every year, varying in type, persons of all classes and belonging to every nation under the sun," Stekel told his audience at the Baden–Baden conference (Stekel, 1950b, p. 281). The increase began before the First World War. During the war, however, there was a decrease in obsessive and compulsive symptoms and a huge revival of hysterical fits. In his consulting-room during the war Stekel could have fancied himself back in Krafft-Ebing's clinic of his student days. Stekel's observation has been confirmed by historians such as Elaine Showalter, who relates the numerous cases of "shell shock" to the overstrained expectations of the manly role in the war (Showalter, 1987, p. 171).

In his Baden–Baden lecture Stekel tried to explain those changes by examining several viewpoints, such as:

- The idea that psychiatric knowledge had expanded, so that psychiatrists diagnosed "compulsive neurosis" where they used to diagnose "hysteria" or "neurasthenia";[5]
- The contention that most cases of hysteria were imitative;

[4] This abridged version does not contain Bien's discussion of the literature on OCN.

[5] Ernst Bien's review of the literature on compulsive processes from roughly 1875 up to 1925 (Bien, 1927–1928) covers more than 2000 studies, two thirds of which were published after 1900.

- The theory that psychoanalysis had unmasked the sexual significance of hysteria, so neurotics had to find an acceptable disguise which would not reveal them;
- The concept that hysteria and OCN succeeded each other from one generation to the next;
- The suggestion that OCN was a post-war disease.

The last suggestion had been defended by Stekel in the concluding chapter of *Zwang und Zweifel* (1927–1928). There, he interpreted the Great War as an alibi to stop reconverting hatred into love, and to give rein to one's inner aggressive tendencies by utilizing them for the fight against the enemy of the fatherland. The war turned hate into a virtue. After the war, however, without this external enemy, people *en masse* locked themselves up in the prison of OCN, where they were kept company by "the inner enemy."[6]

In his Baden–Baden lecture (1930), however, Stekel stated that, though the influence of conditions after the war should not be underestimated, OCN was not merely a post-war disease. The increase in its frequency had begun before the war. In order to account for that phenomenon, the third theory was an alluring one; Stekel rejected it nonetheless, because the vanishing of hysteria could be attributed to the sexual emancipation. An epoch of more sexual freedom had already begun before the war, and after the war "almost all restraints were thrown off." According to Stekel, sexual liberation had been stimulated, but not created by Freud's books. After Krafft-Ebing, Havelock Ellis, Iwan Bloch, and Auguste Forel, the time called for him, he thought (Stekel, 1950b, p. 306).

This sexual revolution was conceived by Stekel as a beneficial change. For many years he had been a fervent protagonist of the battle against sexual bigotism (see Mortier, Colen, & Simon, 1994). However, the relaxation of sexual morality, which began in the upper classes, did not turn out to be the universal preservative for a neurotic development. In his therapeutic practice Stekel had to face the traumatic consequences of a sexual freedom that was not bound up with pedagogical responsibility.[7] In this article we will further explore Stekel's reasons for believing that OCN was a "social disease," a "malady of the time," and more in particular, that "all obsessional parapaths (neurotics) are the abortive offspring of unsound educational methods" (Stekel, 1950b, p. 315).

[6] Stekel (1927–28), Vol. II, pp. 626–632; cf. Stekel (1962a), pp. 608–613. Stekel cynically called the neurotic patient a "criminal without the courage to commit a crime" and the obsessive-compulsive disease the "spiritual prison of the latent criminal."

[7] This is the only Stekelian insight—briefly—referred to in Hale, 1995, pp. 51–52.

OCN and Parental Imperatives

We start with a summary of Gutheil's "preface" and Stekel's "introductory remarks" to *Compulsion and Doubt*. The first compulsions date from early childhood; they originate in the child's inner organs, according to Stekel, who was writing at a time when the "battle of the chamber pot" was a daily phenomenon. The urge to defecate is opposed by a counter-urge. The child may retain his stool for several days. In doing so, the child has made his first stand against cleanliness, the very first objective of training and culture. The stronger the parents insist on daily defecation, the stronger the child's determination to retain his stool. Drive and culture become antipodes. Experience had taught Stekel that most parents did not recognize the ubiquitous character of infantile autoerotic practices. They use threats and punishment to fight masturbation and all other expressions of sexuality. As a result, anxiety appears in the child's life. The ego, pressed by anxiety, accepts the demands of culture, but not completely. In man's psyche a struggle between ego and counter-ego, an antithesis between drive-ego and culture-ego, originates. Disposition and predominantly upbringing determine whether a man achieves an equilibrium between "Pathos" and "Logos" or not.

According to Stekel, OCN ordinarily emerges after the age of five (Stekel, 1966, p. 40). Almost every child displays compulsive symptoms such as grimacing, tics, looking at the clock, compulsive giggling, counting, ritualistic patterns, and reiteration of meaningless words. In most cases these traits will soon disappear. Obsessive-compulsives have a stronger instinctual life than healthy persons have, their intellect is far ahead of their years, and maybe they have a deviant functional organisation of the brain as well (Stekel, 1927–28, Vol. I, p. 110).[8] They have been brought up strictly, all sorts of religious and moral inhibitions have been imposed on them. Inwardly, they do not recognize the educational demands, while a strong inner defiance insists on the demands of the drive-ego, the result being OCN. Obsessive and compulsive symptoms which began as a kind of playful acting (*Schauspielerei*) develop steadily into a *system*, as a rule at the time around puberty (Stekel, 1950b, p. 286). Being ashamed of it, the sufferer is inclined to hide his systematization.

In all compulsion diseases the parental authority is incorporated into the patient's obsessional system. We find it in symbolic disguise, alongside the symbolic representation of instinctual cravings. "Both represent poles of an emotional high-tension system," according to Emil Gutheil in his aforementioned preface.[9] Stekel, who himself had an outspoken sense of

[8] The remark about the brain was omitted in *Compulsion and Doubt*, the English (abridged) edition.

[9] Gutheil, "Preface," p. 16f.

drama, called the compulsive person a comedian; he performs *as if* the parental imperatives from his early life and his reactions to them were the pattern of his emotional life.[10] The obsessive-compulsive is a rebel in disguise, he is inspired with hatred, full of criminal impulses, according to Stekel, who never eschewed robust expressions. In the symptoms one recognizes remainders of his struggle against cultural rules and pedagogical authority—defence reactions to parental imperatives. Some cases appear to have been built around the imperative "hurry up." Not that the patient follows it in his illness; he caricatures it. One part of his personality, representing the parent, urges him to rush, while another part is hopelessly involved in details of executing the intended action, causing endless delay and "slow motion." In their washing mania, these patients rebel against the necessity of keeping physically clean; in their blasphemous and obscene obsessions they rebel against the imperatives of devotion and decency.

We have seen that, according to Stekel, pedagogical imperatives are caricatured in the obsessive-compulsive's system. This interesting idea, already present in Stekel's *Nervöse Leute* (Nervous people, 1911), was further developed in the first volume of his pedagogical letters, the *Briefe an eine Mutter* (1927–1929). It was taken over by an Adlerian, Erwin Wexberg, who wrote in his *Individualpsychologie* (1928) that a child who has been forced to be perfect, takes revenge by absurdizing parental wishes. The pedantic child hits his parents with their own weapons (Wexberg, 1931; Way, 1950).

Stekel must have felt that it remained unclear *why* patients decided for this particular kind of rebelliousness. In his lecture at the Baden-Baden conference he focused on the unresolved basic conflict that motivated the patient to ridicule his upbringing by overstating its effects.

OCN and the Devaluation of the Parental Ideal

In his Baden-Baden lecture, Stekel termed the sufferer from OCN "object-sick," all other neurotics "subject-sick" (Stekel, 1950b, p. 283f). He does not suffer from a disorder of feeling which is directed toward his own inner state like the hysteric patient, but because of an overcharged emotional relationship (a fixation) to another object, who might die or sustain an accident if the patient fails to perform a particular act; Stekel referred to this phenomenon as the "death-or-disaster proviso" (*Todes-oder Pechklausel*). Stekel declared that thirty years of experience had taught him, that the illness was invariably

[10] The conception of the neurotic person as a "comedian" was treated in one of Stekel's earlier popular sketches, collected in *Nervöse Leute* (Stekel, 1911b).

(!) directed against a member of the patient's family. In *Zwang und Zweifel* Stekel still held the conviction that in most cases the "home object" was a brother or a sister (Stekel, 1927–1928, Vol. II, p. 278), in his Baden–Baden lecture (1930) he stipulated that the more usual object was one of the parents. The patient's attitude toward this person is ambivalent; he "invests" the object with both love and hate. The patient's strong affect of doubt—"Did I perform the act correctly?"—corresponds to this bipolar attitude.

In his book on the problems in the woman's love life, *Die Geschlechtskälte der Frau* (Frigidity in Woman), anxiety disorders had been interpreted by Stekel as the possible outcome of an upbringing in which a neurotic mother tries to overcompensate her lack of genuine maternal feeling by an excess of tenderness. In other words, obsessive-compulsives have been mothered by (usually hysterical) women with an ambivalent feeling toward their children (Stekel, 1921b, p. 477). In Stekel's Baden–Baden lecture there is again a "schizoid" family constellation, but its pathogenic influence is not rooted in a morbid interaction structure, it has the character of a *shock*. The "objective parapath" suffers from a shattering of the parental ideal, his childlike belief in the infallibility and essential goodness of father or mother has been shaken.

Looking back upon his many cases of OCN, Stekel concluded "that invariably[!] the parents had failed to practise what they preached to their children" (Stekel, 1950b, p. 299). Here is one of several instances: "A bigoted mother had brought up her son on very strict religious and moral principles, but the son had occasion to notice that his mother on a pilgrimage gave herself to a stranger. The discrepancy between practice and precept in the mother was most disturbing to the child, and completely undermined the latter's authority complex" (Stekel, 1950b, p. 300). This was not an exceptional case, but an everyday tragedy, according to Stekel, who claimed his casebooks being filled with patients who had a mother with a lover or a father with a mistress, or who had been sexually abused by their parents.

Stekel acknowledged the influence of war conditions on the sexual behavior of husbands and wives, fathers and mothers, who had ample opportunities to explore the extramarital domain.[11] When he was working as a military hospital physician, Stekel had personally experienced the disruptive influence of the war on his own marriage and family life. In his autobiography he admitted having had many liaisons with women; with one of his "flames," Hilda Milko, a married woman with two little children, he would develop a permanent relationship (Stekel, 1950a, p. 180f).

[11] Venereal disease, abortion and divorce were the consequences. They were the concern of the *Deutsche Bund für Mutterschutz* (German Alliance for the Protection of Mothers) and its periodical (1908–1932), edited by the well-known feminist Helene Stöcker, *Die neue Generation*, to which Stekel contributed.

Having had more sexual freedom than their parents, the mothers and fathers had "lived their own lives," but their children had been brought up with the old values. They had controled the sexual life of their children and had inculcated moral principles which they themselves no longer practised. In all of these cases the semidivine image of the parent had been shattered. The children found reasons to despise where they were expected to honour, and reasons for hatred where they ought to love (Stekel, 1950b, p. 307). They turned against their parents the traditional morality which these had implanted in the nursery. In doing so, the children took vengeance for the prohibition of sexual amusements. The formula runs as follows: "You have instilled moral principles which were to pilot my life, but you, yourself, do not abide by the principles you teach. Therefore, I condemn you as you would condemn me were I to violate your teachings" (Gutheil, 1962, p. 17).

The traumatic experience of "shattered authority" undermined the child's feeling of security and his sense of self-esteem (the respect felt for educators being an essential element in his self-respect). Many neurotic symptoms, and especially doubt, are direct consequences of the patient's desire to strengthen his uncertain position in life. OCN may be called a "doubt neurosis." "Am I really my parent's child?" This uncertainty is the main root of much morbid doubting, according to Stekel, who also used the Adlerian concept of the "central idea" (*tragende Idee*) to designate the doubter's pressing question (Stekel, 1962a, p. 474). In former times, suspicion of parental adultery was part of the fantasy of jealous children, creating their own "family novel" (*Familienroman*); by now, in many cases there is a profound justification for uncertainty, according to Ernst Bien, writing on the Stekelian view on OCN in 1933 (Bien, 1933, p. 300).

The core conflict is frenetically hid by the patient; it is his "secret" (*Geheimnis*). In his first paper on compulsive states (1910), Stekel, following Freud's first formulation, sought the origin of compulsions in the repression of an unacceptable experience, later he emphatically insisted that the painful experience had not been repressed, but wilfully denied by the so-called *scotomization* mechanism; in other words, the condition is not one of "can't see" but one of "won't see" (Stekel, 1962b, p. 363).[12] The experience has not been forced into the patient's unconscious, but shifted to his day-dreams and notably to his masturbation dreams.[13]

[12] See also the essay on "Nichtsehenwollen" (unwillingness-to-see) in: Stekel, 1924, pp. 13–22. Stekel's mechanism of "scotomization" precedes Sullivan's "selective inattention"; cf. Brown, 1967, p. 187.

[13] Stekel denied the existence of an unconscious with repressed ideas beyond conscious recall; cf. Stekel, 1962a, pp. 225–227. In this respect he declared himself a follower of Freud's French rival Pierre Janet. With his rejection of the Freudian unconscious, Stekel placed himself outside psychoanalysis, according to Hans Christoffel (1939, pp. 668–670) in his critical overview of dissenting schools.

The patient has a "histrionic" nature; he behaves *as if* the traumatic scene has never existed; he plays all sorts of tricks to persuade himself that the event in question is unknown to him (Stekel, 1962a, p. 229). In an attempt to "correct the past," reality has been replaced in the end by a fictive world, or by, what Stekel referred to with a term from the French author André Gide, "a second reality" (Stekel, 1950b, p. 313). What began with a destruction of the parental ideal evolved into a disintegration of the whole authority complex, i.e., of all persons and institutions set in authority over us. Even and notably the supreme authority, God, has been devalued and replaced by the patient's own authority—he is "Autotheos," as one of Stekel's patients had written on the title-page of his diary (Stekel, 1950b, p. 298). An interpretation which reminds us of course of what Karen Horney described as "false pride" and what present-day analysts are used to term "grandiosity" (cf. Salzman, 1980, p. 54).

Realism and Cogency of Stekel's Theory

Stekel's final view on the causation of OCN is extreme, one-sided and all-inclusive. It is beyond doubt that Stekel defended the view that in *all* cases of OCN parents had given reasons for doubting their moral integrity. How did he know? The relevant events were inferred "intuitively" by Stekel from the patient's dreams. We wonder, however, whether Stekel's "transcendent intuitive genius" may sometimes have been guided by his wish to confirm a "fixed idea." For instance, the following claim in his second book on dream interpretation, published in 1935 can hardly be called scientific: "In all cases of obsessional parapathy I have been able to prove that in the patient's youth the authority complex was shattered by the mother's adultery or the father's loose behavior" (Stekel, 1962b, p. 514).

The explanation of OCN as being caused by parental faults can also be found in Ernst Bien's treatise on the subject in the Stekelian handbook for mental guidance, *Der Seelenarzt* (1933). Stekel mentioned his colleagues Pollak, Richter, and Wengraf, who, when dealing with cases of OCN, had "discovered" the same antecedent of this illness (Stekel, 1950b, p. 303). One decade after Stekel's death, two of his former pupils, Gutheil and Lowy, writing on the etiology of OCN, felt obliged to put their master's final and radical vision in perspective—undoubtedly because it could not explain all of their own clinical cases.[14]

[14] Lowy, in his introduction to the revised translation of Stekel's *Technique*, maintained that "very many cases" revealed a breakdown of the patient's belief in the moral authority of his parents. Gutheil, in his introduction to Stekel's *Compulsion and Doubt*, wrote that only "some cases" of OCN can be traced to a shattering of parental authority.

For a fair and balanced understanding of Stekel's position, it is necessary to point to the fact that he neither maintained that every child whose parents break the code is bound to suffer from OCN, nor that an infantile sexual trauma necessarily leads to OCN. In an article on the results of psychoanalytic cures (1913), Stekel indicated that children seem to tolerate sexual traumas much more easily than adults. Even after having experienced numerous traumas, some children remained healthy. Traumas in themselves have no significance, according to Stekel. The trauma is only effective under certain conditions.[15] That is also the case with the bad example of parents. Much depends on the stage of development of the superego and on the level of morality of the environment (Stekel, 1950b, p. 304). The actual situation of the patient is also crucial.

In the aforementioned article Stekel referred to the *exonerative use* a patient makes of a trauma. The patient who cannot cope with an actual life problem (the "precipitating factor" of his illness) tries to resolve it by reviving an old trauma. He displaces the blame for his failure on that trauma (Stekel, 1971, Vol. II, p. 299). This insight brings Stekel close to Jung's view on the significance of the actual conflict (which the patient does not dare to acknowledge) as well as of Adler's teleological interpretation of the concept of "regression" (Ronge, 1934, p. 90).

In *Sadism and Masochism* (1925) Stekel even maintained that "[t]he parapathic constructs for himself the traumata of his life history and his earliest attitudes, just as the heroic peoples make history for themselves" (Stekel, 1968, Vol. I, p. 235). Stekel faced the possibility that the patient reports painful experiences with his parents which in fact are fantasies. After having given up the original "seduction theory," Freud had declared that it is not relevant for therapy whether a trauma has actually happened or not. Stekel's position was not unlike Freud's. What matters is the fact that the patient treats the idea of parental wickedness as a reality (Stekel, 1950b, p. 309f).

The reader may wonder whether here Stekel is minimalizing the importance of real experiences with sexual abuse and pedagogical blunders as causal factors for OCN. As a therapist he was mostly dealing with patients who were adults, and in most cases it was not possible to decide whether their memories were false or not. Stekel's assertion that this did not matter for therapy weakened of course the cogency of his theory on OCN as a "social disease"—that is, as a disease closely linked with an identifiable process of cultural transition, the effect of a witnessed "double sexual standard."

[15] See the article on "The end results of psychoanalytic cures" in Stekel, 1971, Vol. II, pp. 290–321, esp. p. 298. With several modifications this article has been included in Stekel, 1950b, pp. 365–403.

Stekel had been an analyst long enough for the children of former patients to come to him for treatment. In other words, he had opportunities for corroboration. This being the case, it was not so easy to achieve convincing results, as will become evident from the following illustration: A mother had practised fellatio on her son, aged five. Sixteen years later this son consulted Stekel, suffering from severe compulsive problems. Stekel believed that his illness stemmed from the infantile trauma and was interested in knowing if he had any remembrance of it. Alas, what happened? He broke off the analysis "when we began to draw near the repressed truth" (Stekel, 1950b, p. 300). This is an interesting instance, because it clearly reveals Stekel's eagerness for confirming his theories by means of suggestion. His patient may have refused being brainwashed for the sake of an analyst's obsession!

Originality and Influence of Stekel's Final View on OCN

Stekel considered the theory of shattered authority as a very important discovery of his own. Two years after the Baden–Baden congress, in 1932 at a conference in France he proudly presented it as his social theory of the construction of neurosis (Stekel, 1937). But, was Stekel's claim to originality really warranted? Was Stekel merely following Imre Hermann, a disciple of Sándor Ferenczi, who, according to Stekel, had come to the same conclusion in 1929? (Stekel, 1950b, p. 304). However, already in 1925, in his research into sadism and masochism, Stekel had found that the sadomasochist's paraphiliac (= perverse) behaviour was the outcome of an over-compensation for a mother-goddess who turned out to be a whore or for an adored father who turned out to be a Don Juan (Stekel, 1968, Vol. II, p. 431f).

Nevertheless, not all aspects of Stekel's final view on the etiology of OCN were new. Stekel assured his readers that the more opportunities he had to gain insight into the past experiences of neurotics, the more frequently he had uncovered incestuous episodes during childhood (Stekel, 1966, p. 176). In other words, Stekel's theory may partly be interpreted as a regression to or a restoration of Freud's first explanation of OCN, as being a disease caused by a real sexual trauma—the so-called "seduction theory." Jeffrey Masson, in *The Assault on Truth* (1985), pushed Ferenczi forward as the hero who dared to restore—in 1932—the "seduction theory" which had been suppressed by Freud (for a critical discussion of Masson's views, see Stanton, 1991, pp. 105–110). However, we are inclined to value Ferenczi's restoration as a confirmation of Stekel's conviction that often real sexual traumas are at the basis of severe neuroses—an insight which had been expressed some years prior to Ferenczi's "rediscovery."

It has been Stekel's (and also Adler's) fate that several of his observations and suggestions were adopted by other writers without acknowledging their origin. Ellenberger (1970, pp. 638–641) maintained that the striking similarities between ideas of the so-called neo-Freudians (Karen Horney, Harry Stack Sullivan, etc.) and those of Alfred Adler can be ascribed to the latter's influence in the USA. We believe that what they presented as new was also an elaboration of several of Stekel's ideas. Ellenberger referred to the neo-Freudians as neo-Adlerians, but they may be called neo-Stekelians as well.

Given the vast influence of Stekel's writings in the USA, it is surprising to find so few references to him in mainstream psychodynamic investigations. His ideas are not referred to in Leon Salzman's authoritative *Treatment of the Obsessive Personality* (1980). In this work we come across the theory that obsessional patterns develop from a contradictory and hypocritical family background; we quote Salzman's characterization: "When a child grows up in a family where deeds and verbalizations are discrepant and where the child's expectations are not the same as those of the parents, he develops (...) obsessional patterns (...) as a means of coping with the ambivalent feelings that inevitably occur" (Salzman, 1980, p. 49). This theory not only reminds us of Sullivan's view on the pathogenic effect of a "schizoid" family life or of the well-known "double bind" theory (Barton Evans III, 1996, p. 155, 199), but also of Stekel's view on the impact of a family's double moral standard on the child's abnormal development.

Prevention of OCN is Better than Cure

Stekel used to refer to his curative technique as "active analytic psychotherapy" (Harper, 1959, pp. 46–49). What he chiefly aimed at in the active therapy of the obsessive-compulsive illness was "a rectification of the patient's faulty attitude toward the object" (Stekel, 1950b, pp. 308–316). He showed his patients that their symptoms were a regression into childhood. He made them aware of their day-dreams. Clearly, they were reviving an incident from the past. Stekel attacked their "system," and he taught them to use their energy not for seeking vengeance upon an object but for performing self-imposed constructive tasks in the real world (cf. Levenson, 1983, p. 99, who calls therapy "an exercise in self-realization").

Stekel admitted that there was no tougher job than curing OCN because of the patient's stubborn resistances ("will-to-illness"). He is always intellectualizing and attempting to conceal his secrets, especially the more relevant traumatic experiences. Follow-up examinations showed that it was almost impossible to cure an obsessive-compulsive patient completely (Stekel, 1962a, p. 159). Cases where the illness had existed since childhood and especially those cases where real incest had taken place, gave a poor prognosis as to a complete cure. Conclusions like these and the consideration that obsessive-compulsives were lost for society and culture, convinced Stekel of the importance of prophylaxis.

When the illness was caused by the patient's conflict with his parents, resulting from a double moral standard, it was imperative to promote the restoration of a single standard. Stekel saw two alternatives: (a) that parents should practise what they preach, or (b) that children should be brought up in the new moral freedom. In his book on *Sadism and Masochism* (1925), Stekel defended the standpoint that parents have a right to their own sexual life and that children should not pass judgment on their parents. He refused to be "a preacher of morals." Stekel's ideas on this matter come close to those of the Viennese satirist Karl Kraus who promoted in *Die Fackel* (The Torch) the idea of a healthy "living out" of all sexual desires—a position rejected by Freud, who was, according to Juliet Mitchell, liberal in his opposition to repressive bourgeois morality, but antagonistic to what he considered to be symptoms of "sexual corruption" such as promiscuity (Mitchell, 1979, p. 424).

In his Baden–Baden lecture (1930), Stekel thought it very doubtful whether children could be brought up to accept their parent's sexual freedom. The main reason was that one had to reckon with the Oedipus situation and the factor of jealousy: "If the son is naturally jealous of his father, the daughter of her mother, how much stronger will be this passion when it applies to an illicit lover of either parent" (Stekel, 1950b, p. 316). Moreover, official morality (as defended by the Church) will condemn promiscuity. Therefore, Stekel felt obliged to advocate the first alternative.

Parents had to do their best to live up to their principles, their good example should be a trustworthy "object-lesson" for their children. In his pedagogical work entitled *Briefe an eine Mutter* (1927–1929), Stekel warned parents and others in charge of children against taking them to bed or even having them sleep in the parents' bedroom, because of the danger of "improper approaches" and possibly more. Stekel did not confine himself to recommendations such as these, for he emphatically declared that a prerequisite for the success of his alternative was a reform of marriage—a topic discussed in *Marriage at the Crossroads* (Stekel, 1931; see also Groenendijk, 1996, p. 158f).

Obsessive-compulsives were the offspring of unhappy marriages, the witnesses or even victims of conjugal infidelity. Hence, only happily married couples should be entitled to bring children into the world. Procreation should be deferred until the marriage had proven successful. With this suggestion, Stekel, who prided himself on being a freethinker, was deliberately not in tune with the standards of the Church. Even more heretical were his proposals of a trial marriage and the abolition of all impediments to a divorce. Stekel was not naive in believing that "proved happiness" was a lifetime guarantee for conjugal fidelity. Still, a reform of marriage would counteract the epidemic spread of OCN, certainly when the reformed marriage was also a well-informed marriage. For that reason he wrote *Erziehung der Eltern* (The Education of Parents, 1934), a book on the education of (future) parents, in which he tried to convince them of the far-reaching influence of the parental model on children.

Epilogue

Stekel's view on the pathogenesis of OCN as a disease caused by family problems and pedagogical faults is a theory with far-reaching implications. This theory made parents responsible for the psychic invalidism of their offspring, and also for the cultural and socioeconomical unproductiveness of at least some of the new generation. Stekel not only blamed the parents, but also the Catholic Church with its procreation-oriented doctrine of marriage and its meddling divorce laws. He also blamed psychoanalysis—or rather its abuse—for offering certain parents an alibi to follow their own bent in sexual matters, while claiming to give their children sex education. Stekel did not only believe that parents had a good deal to do with the genesis of OCN as a psychiatric affliction, he also reckoned with the possibility that intrusive sexual experiences with the parents—resulting in the shattering of the authority complex—were traceable in the life history of schizophrenics: they fled into insanity because they could not bear the painful truth of incest (Stekel, 1962a, p. 598; cf. Stekel, 1950b, p. 304f).

The "sick nurture" approach to OCN provided "objective" ground for people to explain their problems; individuals also learned to see themselves as victims of pedagogical blunders. The exonerative use of their insights was countered by therapists like Stekel, who followed in this respect Alfred Adler, through a deprecatory way of portraying their patient's lifestyle, emphasizing their irresponsibility, infantilism, etc. (see also Allen, 1959, p. 55–58). Stekel, who conceived of neurosis as a "malady of the time," was evidently aware of the unintended effects his profession might have (in providing an excuse for poor performance); he called OCN an "analytical parapathy" and psychoanalysis a "disease" (Stekel, 1950b, p. 391, 282; cf. Stekel, 1962a, p. 440).

With his perspective on the etiology of grave neuroses and psychoses, Stekel may, in retrospect, be called a forerunner of the orientation of the neo-Freudians, who searched for conflicting cultural trends which might be responsible for individual conflicts (Horney, 1966, pp. 168–182). He helped to pave the way for the family-paradigm which became widespread after the Second World War (Benedetti, 1989, p. 42). For several decades social scientists, educationalists, and therapists, as well as the mass media, cherished the credo ("overcharged idea") of family processes and educational experiences as the determining forces in human development and psychopathology (see, for instance, Ackerman, 1958; a rather influential book, based on the assumption that mental illness is a product of "mental contagion" through family relations). In other words, Stekel's work belongs to the prehistory of an environmentalist approach which has come under heavy attack in recent years (see Gross, 1979; Harris, 1998).

Nowadays, within academic circles a biological approach is very much dominant in analyzing the mechanisms underlying mental disorders (Stone, 1997). It is now generally considered inappropriate to blame the parents for the emergence of OCD. Many expect that in the near future neuropsychological and biochemical research will unveil most of the remaining mysteries of OCD and related disorders (see Wolman & Stricker, 1994; Tallis, 1995). The future will teach us to what extent chemistry or even gen-technology will succeed in minimizing age-old pedagogy as a determinant force in the moulding of the human soul, and in replacing forms of psychotherapy as the obvious agency for the healing of its agonizing problems.

Chapter 5

THE ART OF IMITATION
Wilhelm Stekel's *Lehrjahre*[1]

Jaap Bos and Leendert Groenendijk

Introduction

Having focused on the differences between Freud and Stekel in the previous chapter, here we shall zoom in on some similarities in their discourses. Stekel (1950a, p. 113) claimed that he began practising psychoanalytic psychotherapy in 1903 and that his first "real case"—a 42-year-old rabbi, whose case is described in his book on *Nervous Anxiety States* (1921)—was referred to him by Freud. After his break with Freud in 1912 he continued to call himself a psychoanalyst (although he would refer to the practice as "Psychanalyse")[2] and believed, in the account of Fritz Wittels, that Freud's school had become "entangled in mysticism and philosophy," whereas he himself had remained faithful to "true analysis, the medical work" (Timms, 1995, p. 114).

While far from acknowledging this claim, Freud did admit in a letter to him after their ways had parted that Stekel had remained "loyal to psychoanalysis." However, he added, even if he had been "useful to it" (i.e., psychoanalysis), he had also "done it great harm" (Freud to Stekel, 13 January 1924). It is well known that the break with Stekel was a considerable relief to Freud, and he would come to speak of him in a less than friendly manner, referring to him in public as someone who had gone "totally astray" (1914, p. 19) and, in private, reserving much worse terms for him, such as the characterisations "morally insane" (Freud to Ferenczi, 13 September 1912, cited in Jones, 1953–1957, Vol. II, p. 137) and "imbecile" (*schwachsinnig*) (unpublished

[1] This chapter first appeared in *The International Journal of Psychoanalysis, 85*(2004), 713–730.

[2] The only other analyst who referred to it as "Psychanalyse" was the Swiss analyst Oskar Pfister, who believed this to be a linguistically more correct spelling.

letter of Freud to Rank, 13 September 1912). Moreover, Freud had always been reluctant to acknowledge publicly Stekel's position within the "movement," he rarely hinted at their one-time close collaboration and, never in the 50 or so references to Stekel throughout the 24 volumes of the *Standard Edition* did he ever refer to him as his (former) "follower." Instead, Stekel is referred to as "a colleague" or "an experienced colleague" or, more often, simply as "Dr Stekel" or "Wilhelm Stekel," a fact that had not escaped Stekel's attention: "Freud quoted me repeatedly 'Herr Dr Stekel,' which otherwise with other authors (Adler etc.) was never the case" (1926, p. 553).[3] His defection is only dealt with in passing in the polemical "On the History of the Psychoanalytic Movement" (Freud, 1914), as if it were an entirely insignificant event. Only in 1925 would Freud publicly say something about Stekel's former capacity as a psychoanalyst, admitting even that he himself had at one time been "under his influence." In a passage added to chapter 5 of the 6th edition of *Psychopathology of Everyday Life*, Freud wrote:

> That writer [Stekel], who has perhaps damaged psychoanalysis as much as he has benefited it, brought forward a large number of unsuspected translations of symbols; to begin with they were met with scepticism, but later they were for the most part confirmed and had to be accepted ... Stekel arrived at his interpretations of symbols by way of intuition, thanks to a peculiar gift for the direct understanding of them. But the existence of such a gift cannot be counted upon generally, its effectiveness is exempt from all criticism and consequently its findings have no claim to credibility (Freud, 1901, p. 350).

A man with only an intuitive understanding of dream symbolism who had little patience with scientific enquiry and whose work was a nuisance to science *because* it is so often correct: by and large, this is the historical image that has stuck to Stekel (Sachs, 1945; Jones, 1953–1957; Roazen, 1975). Even Fritz Wittels, who sided with Stekel's camp in his conflict with Freud (but later switched sides) and consequently came to know him fairly well, reflects this image, writing of Stekel that he was a "dream interpreter whose equal I have never seen" (cited in Timms, 1995, p. 112), yet scientifically he seemed to have been "like the mental arithmeticians who call out the results without using any observable method" (ibidem, p. 114). Indeed, "Stekel's method was not to have any" (ibidem, p. 115).

In this chapter we shall challenge the assertion that Stekel was not a "true follower" of Freud because he lacked scientific integrity and had a different (faulty) understanding of psychoanalysis.[4] Focusing on the autobiographical element in Stekel's publications, we aim to show that, to all intents and purposes, Stekel's work was much more in line with that of Freud than is commonly believed; more specifically, we shall claim that Stekel's method of dream interpretation was copied from Freud's in several respects, and that this imitation contributed to his distancing from Freud.

A Brief Overview of Stekel's Main Works

Stekel's publications date back as far as 1895 when his first medical paper appeared—a study on "coitus in childhood"—which was cited by Freud in 1896. However, it was not until around 1901 that he became personally involved with psychoanalysis when he read *The Interpretation of Dreams* (1900) and went into brief therapy with Freud. Consequently, Stekel began to devote much of his time and energy to propagating psychoanalysis (for example Stekel, 1907a). By 1912, the year he broke with Freud, he had written a considerable number of newspaper articles and booklets on psychoanalysis as well as four larger volumes on this subject: *Nervöse Angstzustände* (a clinical book on anxiety neurosis), *Dichtung und Neurose* (on the psychology of art and artists), *Die Sprache des Traumes* (on dream symbolism), and *Die Träume der Dichter* (on the psychology of poetry and crime). Apart from these works, he had also published several books on loosely related subjects, such as *Nervöse Leute* (1911b, a collection of popular tales on psychology) and *Äskulap als Harlekin* (1911a, a popular book with humorous sketches from the practice of a medical doctor, in which, however, neither Freud nor psychoanalysis are mentioned).

Sure enough, by this time Stekel was considered to be the most important psychoanalytic author after Freud. His book on anxiety neurosis (Stekel, 1921a)[5] drew much attention, as can be gathered from a letter from Jung to Freud dated 19 June 1908 (Freud & Jung, 1974, p. 157), and it probably won many physicians over to psychoanalysis. His influence on the early reception of psychoanalysis is easily underestimated.

After his break with Freud, Stekel continued to practise psychoanalysis, although his modification of it markedly differed from that of the Freudians: for one thing, he was one of the first to practise short-term therapy, which greatly enhanced his "cultural authority" in the United States (Park, 2004). For another, he cared much less for transference during therapy and, in fact, only accepted analysands if he could get along with them on personal terms (Stekel, 1950a, p. 117). Stekel also remained as productive as before. His most important post-1912 publication is the large ten-volume series "Disorders of the Instincts and the Emotions," published between 1912 and 1928, his

[5] We refer to the 3rd, much enlarged edition of *Nervöse Angstzustände*; the first book edition appeared in 1908. An even earlier version was published as an article (Stekel, 1907a).

magnum opus. All of these volumes have been translated into English. Books on dream interpretation (1962b) and technique in analytical psychotherapy (1950b) are among his last published works. His autobiography (1950a) appeared posthumously.

While it is difficult to give a precise estimate of the total number of publications by Stekel, it is easier to characterise his work in general. (In addition to almost 40 books and booklets, Grinstein (1958) lists some 300 articles that can be attributed to his name; but recently Clark-Lowes (2001) claims to have identified well over 500 articles.) In contrast to Freud, Stekel always remained a clinician first. His "Disorders" series is, in fact, a huge project that focuses exclusively on the description of case material. The literally hundreds of cases described therein are virtually void of the sort of metapsychological speculation Freud was fond of. But Stekel had always and would always content himself with a simple theoretical framework and abhorred all the finer distinctions Freud made, and was consequently often criticised for "confusing concepts" and "jumbling up everything" during the discussions of the Vienna Psychoanalytic Society (*Minutes*, Vol. I, pp. 178–179, 208, Vol. II, p. 7).

Of all these books and articles, only a few are referred to by Freud. Apart from the above-mentioned paper on coitus in childhood, Freud explicitly mentioned the following six sources: a 1904 newspaper article on "unconscious confessions" (a discussion of which takes up two full pages in the 1909 edition of *The Psychopathology of Everyday Life*); his 1908 book on "anxiety neurosis" (in three different places); a 1909 paper on dream symbolism, and the large 1911 book on the same subject (discussed in *The Interpretation of Dreams* as well as in several papers); and two 1911 papers that were published in the *Zentralblatt* (in two technical papers, see Freud, 1912a,c).

This may seem a meagre score considering Stekel's enormous output but, compared to references Freud made to other followers, and considering the fact that there are many more references to Stekel without a specific source indication, one should actually conclude that Stekel occupied no insignificant place in Freud's discourse. We return to this observation in the last part of our paper, but shall first focus on those features of Stekel's discourse that provide the strongest links to that of Freud—the autobiographical narrative.

The Autobiographical Stekel

Freud's first step into the new sciences that was to become psychoanalysis was, as is well known, his famous "self-analysis," and the two books in which this self-analysis is explored most thoroughly and carefully are *The Interpretation of Dreams* and *The Psychopathology of Everyday Life*. The accounts in these books about Freud's life are presented to the reader half as case histories and half as autobiographies. That is to say, in both books, the personal is very much intertwined with the epistemological part of the story: to understand a particular (symbolic) element in the narrative, intimate knowledge of the narrator's life is required. However, that in itself is not enough; one must also understand the generic logic that lies beneath the symptom formation. The play of interpretation therefore consists in connecting the personal and the general in such a way that the target of the interpretation (the narrator) cannot deny its validity. This is what has come to be known by Grünbaum (1984) as the "tally argument": the fact that an interpretation must "tally with what is real" in the unconscious. There is, however, an unintended consequence to this tying of the self to its own accounts, which could be called the autobiographical trap: all accounts lose their innocence and are forever referred back to the autobiographical self since no one can escape his self. Consequently, ever since Freud it has become common practice to demand from people that they reveal their autobiographical self in their accounts.

The Interpretation of Dreams and *The Psychopathology of Everyday Life* had only just been published when Stekel first met Freud, and they found a fruitful interpreter in him.[6] He was not only quick to understand that there are always hidden symbolic meanings behind every narrative, but he was also quick to draw the conclusion that nobody could escape the autobiographical trap, not even—or perhaps, in particular, not—the master himself. When Freud presented a dream of his own to the Vienna Psychoanalytic Society, Stekel was the only one who dared allow the dream to point back to Freud—and he was not contradicted by him either (see *Minutes*, Vol. II, pp. 180–187, meeting of 1 January 1911). He was perhaps the first in what has become a whole industry of discovering Freud beneath Freud. However, the autobiographical trap has implications for the position of the interpreter as well: he, too, cannot escape his own discourses. If a critique of Freud is not to bounce back immediately, it should be levelled from a point of view that is at least equal to that of Freud. And this is precisely what Stekel aimed to achieve, as we shall attempt to show below. However, in our reconstruction of Stekel's position we aim to stay clear, as far as possible, from *hineininterpretieren* and instead intend to draw parallels between the discourses of Freud and Stekel based on rhetorical properties of those discourses.

In what ways, then, can Stekel's and Freud's discourses be compared? First, much as with Freud in the famous *aliquis* case in *The Psychopathology of Everyday Life*, Stekel, too, often presents his case histories in a novel-like style, frequently using the first person singular to create an immediate

[6] Stekel had a copy of both books with a dedication by Freud acknowledging their friendship and, above all, his early appreciation of psychoanalysis—Stekel wrote a positive book review of *The Interpretation of Dreams* in 1902—(see chapter 8).

understanding with his audience. In large chunks of dialogue Stekel makes his appearances as a Socratic interviewer; he knows all and almost effortlessly directs the patient to his unconscious drives. The patients in these narratives, in turn, seem focused very much on him as well, often misspelling or forgetting his name and thereby betraying his real intents, or otherwise dreaming about him. However, unlike Freud's examples, Stekel's cases never leave room for doubt. They are always complete successes. Freud somewhat cynically wrote of Stekel's book on dream symbolism, "The aura of optimism that hovers over the whole performance—we are always right, all our findings fit together, are useful, and so on—may repel us, but the practitioner won't mind it at all; they welcome illusions" (Freud to Jung, 13 August 1908).

It is true that many of Stekel's cases seem to fit too perfectly, as if they were made up. They are sometimes so crude, even to the point of being fantastic, that they appear to be almost a parody of psychoanalysis. It is well known that some analysts (in particular Tausk) thought that Stekel made up his own case histories. Stekel, who was well aware of this accusation, somewhat clumsily replied, "If I had invented my cases I should undoubtedly be a greater poet than Shakespeare" (1950a, p. 142). His proverbial "Wednesday evening patient" has nevertheless continued to be a testimony to his unscrupulous unscientific methods. It may well be, however, that the impression of forgery is actually prompted by Stekel's attempt to imitate Freud's narrative style. The following case history, one of many, may serve as an example: a patient dreams about *being with a doctor, even as she is about to leave. While dressing herself, some slimy substance drips from her hand on to a piece of paper.* "Of course *I* am the doctor," explains Stekel, and "dressing herself is the reverse of undressing herself," and yes, "the slimy substance is sperm. It's about coitus" (Stekel, 1921a, pp. 323–324).

Such self-glorifying narratives obviously made Stekel even more vulnerable to critique than already was the case. Moreover, it becomes even worse when the rivalry between Stekel and Freud surfaces in the dreams of his patients, as happens in *Die Sprache des Traumes*, where a patient dreams of *going up on a ladder* and shortly afterward dreams of *a professor*. The patient, who is a psychoanalyst himself, "resists the uncovering of his inner secrets;" he wants to "complain about Stekel to Freud," says Stekel. "He wants to stand above me, he wants to triumph over me with the help of Professor Freud" (1922a, p. 393).

His rivalry with Freud had already surfaced during the meetings of the Vienna Society, but the autobiographical trap made it almost impossible for Stekel to criticise the master openly. Consequently, some of the more critical narratives found their way into the "popular works," where they are covered up with humour or a pseudonym (or both). His 1911 *Äskulap als Harlekin*, for example, contains a barely concealed complaint about Freud appropriating his ideas. This book, which was published under a pseudonym, appeared at a time when he had made public two of his most important findings: "the law of bipolarity" (the hypothesis that every act is determined by two opposing forces) and the discovery that all dreams are "polymorph criminal" (*universell kriminell*), a rival to Freud's hypothesis of the "polymorph perverse unconsciousness." Both ideas were unanimously denounced at the meetings of the Vienna Psychoanalytic Society. Freud is quoted as saying, for example, that the concept of the "criminal instinct" does not explain anything, that it is the viewpoint of an outsider (*Minutes*, Vol. III, min. 155), and others were quick to reduce the "law of bipolarity" to a mere rephrasing of what the Professor had already said. Interestingly, from Lou Andreas-Salomé's diary notes (1987, p. 94), we learn that it seems as if Freud did not think so low of the concept of "polymorphous criminality," which, although considered by him to be a "methodological leap," apparently hit a deeper meaning: the anxiety about oneself. Stekel himself complained of how they made fun of him in the society, nicknaming him "Stekel with his Bipolarity" (1950a, p. 132). The disappointment he felt about this hostile reception is reflected in the first story in his *Äskulap* book, which is about a humorist who desperately wants to have an idea of his own. When he finally finds one, he bumps into "his old professor," bureaucrat X: "The Professor assures me I am a real original thinker (what irony!). He reads my stories with great pleasure. I ought to write something nice again real soon." When he leaves the "Professor," the idea is mysteriously gone: "Is it that between one's own thoughts and bureaucracy professors stand, or is there between these ideas and professors who are bureaucratic, an enmity?" (Stekel, 1911a, p. 4). The story ends with the humorist being committed to an asylum, where his case is presented to an assistant with the following speech: "an interesting case of monomania ... This man suffers from the newest illusion of grandeur. He believes he owns an idea of his own" (ibidem, p. 6).

A third sense in which Stekel is present in his own work is where he relates his own dreams, slips, mistakes etc. Again, Freud, who with the famous Irma dream in *The Interpretation of Dreams*, opened the door to a whole array of autobiographical pseudoconfessions, sets the example. In the above-mentioned 1904 newspaper article, for instance, Stekel narrates several examples of unconsciously determined actions: he accidentally expresses the exact opposite of what he should have said ("I hope you will *not* leave your bed soon," to a patient) or clumsily unfastens the dress of a lady friend etc.; examples that support Freud's thesis as developed in *The Psychopathology of Everyday Life* and were, for that reason, quoted by him (Freud, 1901, pp. 68–69). More autobiographical material appears in *Die Sprache des Traumes* (Stekel, 1922a, pp. 341–342), where he links the numbers 14 and 41 in a dream with a fear of masturbation (he was 14 years of age when he read a popular book by Dr. Retau warning its readers against this practice, which upset him terribly). The theme of the fear of masturbation returns time and again in his works, as, for example, in his series of educational letters to a mother (Stekel, 1927–1929), where he again relates of his reading Dr. Retau's book and the terrible consequences it had. Masturbation was also the last topic to be discussed in the Vienna Psychoanalytic Society before Stekel was expelled from this society (*Minutes*, Vol. II). Against Freud's opinion that adult masturbation is harmful in a psychological as well as a physical sense, Stekel (1950c) strongly advocated this practice, arguing that those who *repress* the urge to masturbate become neurotic (see chapter 3 for a discussion of the differences between Freud and Stekel on the topic of masturbation).

The implicit suggestion in the masturbation debate is that Stekel himself is the living proof of his thesis. Or, in other words, that there is a fourth set of self-narratives, which are the autobiographical narratives in disguise. Thanks to the publication of his *Autobiography* (Stekel, 1950a), it is now easy to see that a number of case histories published by Stekel are really covert autobiographical accounts. Particularly, his book on dream interpretations (1926b) contains many accounts of covert autobiographical narratives, but even his very first article on "coitus in childhood" (1895) can be reconstructed as a narrative about his own childhood, as the editor of his *Autobiography* already noted in a footnote (Stekel, 1950a, p. 33). In many of these narratives, masturbation again plays an important role. In the next section of this chapter we shall explore one such case in detail; suffice it for now to observe that here, too, Stekel had followed the example Freud set in *The Interpretation of Dreams* and *The Psychopathology of Everyday Life*, which are, as is well known, full of self-narratives as well as covert autobiographical accounts. Perhaps Freud's readers, including Stekel, were more aware of that fact than we realise.

The discussion of the autobiographical position therefore makes it clear that, in order to become an analyst, Stekel strongly identified with Freud and followed him in detail (see chapter 2 for additional evidence to support this claim). However, his identification with Freud's style is only the first step in the transformation of his identity. As we shall explore, the next step consists in the appropriation of Freud's self-analysis.

Confessions and Conjectures: The Case of a Lost Flute

When Stekel wrote his *Autobiography* in London during the last months of his life, he aimed to sketch a picture of his inner drives in an open and unreserved way, so that future generations of students could use it "as a source for analytic research" (1950a, p. 12, editor's preface). He believed to be more or less following the example set by Rousseau, who had spoken freely and unreservedly about his own sexuality (p. 29). Indeed, Rousseau seemed "obsessed with publicising unsavoury incidents, chiefly sordid sexual scenes," says Peter Gay (1995, p. 107).

Interestingly, Stekel had analysed Rousseau's *Confessions* extensively in his book on psychosexual infantilism (1922b). His main argument therein was that, thanks to Rousseau's "infantile exhibitionism," he could uncover the deeper layers of the artistic drive in general, which are made up of a specific combination of homosexuality and paranoia. Unlike the average neurotic, the artist himself transcends these "darker drives" and becomes a creative human being.

If Stekel had meant to follow Rousseau's example, it is remarkable, to say the least, that none of the elements he found in his predecessor surface in his own work. Stekel's "confessions" are neither candid nor unreserved. There is no reference to any overt or covert homosexual drive (in fact, there is a heavy emphasis on heterosexuality) but, more importantly, the autobiography certainly does not leave the impression of a man who was able to transcend the average. Instead, we engage with someone who apparently suffered from delusions of grandeur, and who had not mastered his "complexes," that is, his lifelong struggle with Freud.[7]

In fact, the *Autobiography* does not in the least resemble a confession. It is probably more accurate to say that it was modeled on Goethe's 1795 novel *Wilhelm Meister's Apprenticeship* (1995), about a man's search for a vocation. This book was written as an unusual mixture of autobiography and art—*Wahrheit und Dichtung*, to use the old master's own words. It is for this reason that historians of psychoanalysis generally dismissed Stekel's *Autobiography* as a one-sided, untrustworthy account. Yet, with all its faults and inaccuracies, it does provide clues to Stekel's private life and, more importantly, it allows us to reread his work in a new light, one that in the end also contributes to a better understanding of Freud's influence, we believe. We restrict our reading of this work to one story.

The *Autobiography* opens with an account of his youth in Boyan, Bukovina. Stekel outlines his first recollections, which seem "indifferent, without emotion and without importance" (1950a, p. 31). Such a disclaimer, as any reader of Stekel should know, must of course be taken for the exact opposite—and we shall see why. Stekel remembers the house he lived in, located at a crossroads, and he sees a simple cart in which his grandmother is sitting. A second memory: his grandmother has died and Stekel's mother goes to her funeral. Between these two events, the author situates an event that "determined my whole life." He is about two and a half years old and he visits his grandmother in Boyan. While walking on the street, he meets a girl who asks him to play with her. Off they go and play

[7] The *Autobiography* was still incomplete at the time Stekel died and it is difficult to assess how much the editor altered or changed the manuscript to complete it. Stekel's son, Eric-Paul Stekel, however, expressed considerable doubt about its reliability (see Timms, 1996, p. 48fn.).

> the favourite game of children, 'father and mother' ... Nearby is an improvised shed, which I distinctly remember ... We entered the shed and looked cautiously around us. Then we continued to play 'father and mother', and this time we enjoyed the physical side of our 'marriage' (ibidem, p. 32).

Thirty years later Stekel meets the girl again, now as his patient. She has completely forgotten about their playing together, all she remembers is how his brother once knocked her down: a "screen memory," Stekel concludes. Did he see her a second time? he asks himself. He does not know; all he remembers is seeing himself, sitting in a cart, riding home.

> I have made a small wooden flute like those which peasants make. I try to play on it. The flute falls from my hand to the road. I cry bitterly. The carriage stops. Some of the men go down to look for the flute. The passengers are in a hurry. They shout to the driver 'Go on! Go on!'. The cart rumbles and creaks over the dusty road. I sadly look back. My crying has been in vain. The hot sun presses the tears against my cheeks. The wonderful flute is lost for ever (ibidem, p. 33).

We have quoted considerably from this peculiar mixture of dreamlike recollection and autobiography because it corresponds largely with a case history narrated in the first volume of *Impotence in the Male* (Stekel, 1971). The case history, presented to us as an important *document humain*, is about a 40-year-old medical doctor who goes by the initials of "NM." In this case history, the subject remembers how he, at the age of 3, played "being married" with a girl. They creep into a hayloft and he looks at her genitals, has a strong erection and actually attempts coitus.

> I do not recall whether I repeated this game with the girl. I believe I only visited my grandmother, who lived in the country, for a day, and had to return to the city the next day. This recollection is mixed up with a second one, which I consider a screen memory. I received a cheap shepard's flute from my grandmother, and let it fall from the wagon while we were rapidly driving to the city. The wagon stopped; an unsuccessful search was made for it; and I cried bitterly. The lost flute! Is it not a symbol for the abandonment of the pretty girl, the first of my long series of erotic adventures? (ibidem, p. 117).

A few details do not match exactly: Stekel is two and a half years old, "NM" 3 years of age; the shed is a hayloft; the girl's "screen memory" is now his own, but, even so, these two accounts are clearly identical and the identity is positively established by the "insignificant recollection" of the grandmother sitting in the cart that encases the "life-determining" event. "NM" can be none other than Stekel himself.[8] This fact is not insignificant precisely

[8] There is, of course, another possibility, namely that Stekel appropriated NM's story and made it his own, rather than the other way around. We believe this to be unlikely, however, in the light of other biographical information given about NM, not discussed here. Another indication that the letters "NM" stand for Stekel is that, in Latin, these two letters mean *Nomen Meum* ("my name"). In either case (whether it is truly his story or only a story he "borrowed" to make his own), our reading of it as an account of an auto-analysis remains the same.

because the case history of NM does not end here, but, in fact, gives detailed information about its subject, in particular about his sex life. We thus come to know the hero of the story as a sexually active person who, at a very early age, had already had very intimate relations and later even becomes something of a sexual athlete. What remains vague in the autobiography is clearly expressed in the case history: the "father and mother" game means intercourse. It continues: by the time the subject goes to Gymnasium he practises masturbation on a daily basis; aged 14 he visits a brothel for the first time; and at the age of 17 he has an affair with a chambermaid, which unfortunately results in hysterical impotence. After two years of suffering he meets the woman who will later become his wife; this relationship cured him of his impotence and he was now "able to carry out coitus seven times and subsequently ten times in a brief interval of time," an observation that could be "confirmed by his wife" (ibidem, p. 126).

The basic psychological model behind this narrative is the following: a healthy sexual condition is spoiled by feelings of doubt and fear induced by an unnatural, repressive voice (warning the child that masturbation ruins one's health), which only later results in impotence. Health is only restored when "insight" is gained, allowing the adult to become sexually free again.

Indeed, in the covert autobiographical case history we encounter a strong, self-confident Stekel, athletically built, a free-thinker and a confirmed atheist, free of complexes. "I aspire to the high plane of a *Nietzsche* who believes every person has his own morals," he has NM proclaiming. "But my inner (unconscious) morals are stronger than my intellect. In the course of the years, I have learned to listen to this inner voice, and, since I have done it, know that I will never be impotent again" (ibidem).

Again and again, Stekel would replicate this successful model in the narratives of his patients, whose neurotic problems were often traced back to the repression of masturbation or some other psychological conflict during childhood, and were completely cured after only a brief but "active" therapy with Stekel.[9]

The story of the lost flute is presented to the reader without virtually any interpretation, a peculiarity in itself for a writer who was quick to recognise a symbolic meaning behind almost everything. Thus in the *Autobiography* the story ends on the sad note that "the wonderful flute was

[9] When, in 1925, the Ufa film company proposed making a film about psychoanalysis, Freud flatly refused to cooperate. The film was then made with the aid of Hanns Sachs and Karl Abraham and released as *Geheimnisse einer Seele* (Secrets of a Soul) in 1926 under the directorship of G.W. Pabst. The movie is about a man who becomes impotent when a childhood conflict reemerges later in his life. He is completely cured during what appears to be active psychotherapy. Although Freud did not see the film script beforehand, nor was it likely that he ever saw the completed movie, it was the Stekelian-type of vulgarisation of psychoanalysis that put him off, as he wrote to Abraham that he did not think it was possible to represent psychoanalytic abstraction in any respectable way on film (Freud to Abraham, 9 June 1923, Freud & Abraham, 1965).

lost for ever" (1950a, p. 33). But in the case history of NM, the account breaks off at the point where the narrator "knows that [he] will never be impotent again," whereupon Stekel (1971, Vol. I, p. 117) concludes with the rhetorical question: "Is [the flute] not a symbol for the abandonment of the pretty girl, the first of my long series of erotic adventures?". Considering the fact, however, that the "patient" was impotent for over two years and that he regains his potency only when he returns to a situation that closely resembles the childlike situation of his youth, it is difficult to see how the "lost flute" can be anything but a symbol for precisely this lost potency. However, Stekel purposely chose not to interpret his covert autobiographical account in psychoanalytical terms; he wanted it to remain "blind." In fact, "in simple cases," Stekel wrote, "recovery may be induced by nothing more than suggestion without going into the deeper motives" (ibidem, p. 127).

Stekel's Blind Spots: The Analysis that Failed

If the case history of NM is really a covert autobiographical narrative, as we have argued, then why is it so uninsightful; why does it not clarify its subject's psychology? The answer to this question can be found if we contrast Stekel's methodological viewpoint with those of Freud.

From the autobiographical case of NM, we learn that Stekel was convinced that it is an analyst's duty to actively interfere in the patient's life: "The psychotherapist must ... convince the patient that fear makes him impotent" (1971, Vol. I, p. 127). In order to do so, he should have an immediate understanding of the patient's needs and troubles. He believed that he himself was luckily endowed with the gift of interpretation, as not even Freud would deny. "I think it will be generally acknowledged that I have some skill as an interpreter of dreams," he once said (1962b, p. 71). To him, psychoanalysis resembles an art; an art for which one needs to be equipped with a certain sensitivity to learn it, which not everybody has: "It's an artist's assignment and it can never become a trade" (Stekel, 1922a, p. 442).

Having said that, he acknowledged at the same time that every analyst has his own complexes which psychoanalysis does not clarify to him if he is not conscious of them. I call this the "psychoanalytic scotoma" (blind spots). It is therefore necessary to learn from one's own dream analyses and to accept oneself in the first place (ibidem, p. 442).

These scotoma make analysts "BLIND TO SUCH COMPLEXES AS HE HIMSELF HAS," he wrote in capital letters in his book on dream interpretation (Stekel, 1962b, p. 511), because "every psychotherapist who is under the dominion of an overcharged idea will tend to introduce this idea into any dream he analyses" (ibidem, p. 512).

The key problem of blind spots calls forth the question of training, which, as is well known, began to surface in psychoanalysis for the first time at about the time that Stekel was expelled from the movement in 1912, and which was a question that was not to be resolved until after the Second World War (see Wallerstein, 1998, for a history of the debates on this issue). It is generally believed that it is precisely at this point when psychoanalysis began to professionalise that unorthodox analysts such as Stekel, whose "wild" and "intuitive" methods made him unfit for the profession, were finally done away with. It will come as a surprise, then, that Stekel's position in the question of training analysis was actually quite sophisticated and, in many respects, in accordance with Freud's beliefs on the matter—albeit not in *all* respects.

Building on the assumption of "blind spots," Stekel argued first of all, in a section on "autoanalysis" in his book on dream interpretation, that it is not sufficient for an analyst to analyse his own dreams:

> When my pupils ask me whether auto-analysis is possible, I usually say: 'It is as impossible for a man to analyse himself as it is for a man to play chess against himself. One cannot at the same time be accusing counsel, judge, defending counsel, and reporter. Every attempt at auto-analysis leads swiftly to a frontier which the would-be auto-analyst cannot cross' (Stekel, 1962b, p. 184).

Consequently, he argued, it is imperative that

> every analyst should himself have been analysed, if only as a preliminary to his practising dream interpretation. Of course this would only be a perfect safeguard if an analysis could be relied upon to free the analysand from complexes. But that, alas, is too much to expect. A psychotherapist who is unduly self-centered will not give the requisite attention to the dream his patient relates. Successful intuitive interpretation presupposes that the interpreter can achieve full imaginative insight into the working of the patient's mind. One who is himself preoccupied with cares, doubts, shattering experiences, or is subject to a parapathy [neurosis], may fail to interpret a dream because he is incapable of imaginative insight (empathy) (ibidem, p. 512).

For this reason, Stekel also warned against turning former patients into analysts, and fiercely opposed lay analysis. In his book *Technique of Analytical Psychotherapy*, Stekel (1950b, p. 390) wrote that it is a "mistake" when an analyst educates his ex-patients to become psychoanalysts: "It is obvious that this is the path taken by many of those doctors who become psychoanalysts".

In short, psychoanalysis was to Stekel not a trade you could learn by reading Freud's books, but an art that one could only learn through a proper analysis of oneself, analogous to the master–pupil apprenticeship of old, restricted to those few who have an innate sensitivity for it and who are medically trained. In other words, in all but the last point did he concur with Freud.

How did Stekel himself fit into this model? We remember that he had, indeed, been (briefly) in analysis with Freud around the turn of the century and became an analyst himself shortly afterward. He wrote that it was from Freud that he learned "the art of dream interpretation" (Stekel, 1962b, p. 180). Nevertheless, recall that it was sexual problems (*Wirrnisse*, or "confusions," as he wrote in a 1926 account) that first brought him to Freud. Freud diagnosed these "confusions" to be related to a "mother fixation." Stekel wrote, "I was astonished. My relations to my mother are like those of a normal human being, I was able to separate from her early" (see chapter 8).

So it was Stekel who broke off his therapy with Freud, allegedly after a mere eight sessions. However, the reasons why he did so are not entirely clear. He wrote of this unsuccessful analysis in the *Autobiography*, claiming that he disagreed with Freud on the diagnosis of "mother fixation." At this point, however, the narrative of his analysis abruptly breaks off, only to be continued with an account of how he fell ill, went to Abbazia in Italy, and how he recovered there without any help, *simply by ignoring his doctor's advice*. "What did you do to produce this miracle?" the doctor asked. "I confessed that I had not followed his instructions, that I had taken long walks. He repeated his warning, but I knew that my treatment was concluded" (Stekel, 1950a, p. 110).

There is a threefold suggestion in these parallel stories, which is important for our understanding of both Stekel and Freud. In the way we understand it, the implication would be that Stekel was first of all properly trained by Freud, while he was at the same time cured of his "sexual problems," which had never been psychological "complexes" in the first place but mere "confusions," and lastly, and most importantly, that he had equalled the master because he understood better than Freud his inner psychological condition. Indeed, Freud had *never* been analysed himself and, as a result of that, "Freud, in his *Interpretation of Dreams*, overlooks important complexes where his own dreams are concerned" (Stekel, 1962b, p. 180).

With this conclusion in mind, we return to the lost-flute narrative, which we can now read as an imitation of Freud's self-analysis in *The Interpretation of Dreams*. Much as Freud presented fragments from this autoanalysis but nevertheless did not reveal the Freud behind Freud (the unconscious Freud) and always remains a vague, "opaque person" (*undurchsichtige Persönlichkeit*) in his own work, as Jaspers (1948, p. 646) aptly wrote, suggesting that none of his dreams reveals anything remotely neurotic, so does the covert autobiography of Stekel present its subject as an essentially sane person, free of serious complexes, without "blind spots."

The idea of "scotoma" or blind spots in relation to training analysis builds the last and perhaps most interesting link with Freud. In 1912, after Stekel's book on the *Language of Dreams* was published, in which the idea of "blind spots" first surfaced, Freud wrote a short paper entitled "Recommendations on analytic technique" (1912a). It was first published in the *Zentralblatt*, edited by Stekel, and it contains several rather disconnected sections on technical rules in analysis. One of them (section "f") is on the question of auto- and training-analysis. Freud writes:

> Just as the patient must relate everything that his self-observation can detect, and keep back all the logical and affective objections that seek to induce him to make a selection from among them, so the doctor must put himself in a position to make use of everything he is told for the purpose of interpretation and of recognising the concealed unconscious material without substituting a censorship of his own for the selection that the patient has forgone. To put it in a formula: he must turn his own unconscious like a receptive organ toward the transmitting unconscious of the patient ... But if the doctor is to be in a position to use his unconscious in this way as an instrument in the analysis, he must himself fulfil one psychological condition to a high degree. He may not tolerate any resistances in himself which hold back from his consciousness what has been perceived by his unconscious; otherwise he would introduce into analysis a new species of selection and distortion which would be far more detrimental than that resulting from concentration of conscious attention. It is not enough for him that he should be an approximately normal person. It may be insisted, rather, that he should have undergone a psychoanalytic purification and have become aware of those complexes of his own which would be apt to interfere with his grasp of what the patient tells him. There can be no reasonable doubt about the disqualifying effect of such defects in the doctor; every unresolved repression in him constitutes what has been aptly described by Stekel as a blind spot in his analytic perception (Freud, 1912a, pp. 115–116).

Freud's argument in favor of training analysis and contra auto-analysis seems in perfect accordance with Stekel's position as outlined above. However, it is not too difficult to see how this particular section can be read as a covert critique of Stekel. In the first part, Freud emphasises how the analyst must use his unconsciousness "like a receptive organ" in order to listen to the patients' unconsciousness, and we recall that it was Stekel who was famous for having such an intuitive understanding of psychoanalysis. When Freud argues that it is not enough that the doctor "approximates a normal person," he may have had in mind the sort of critique he vented to Jung when he wrote that Stekel "contents himself with approximations" (Freud to Jung, 13 August 1908). The critical element in these passages is, however, located in Freud's ironic use of the term "blind spots," when he writes that the control of conscious thinking is achieved in training analysis, so as to learn to recognise one's unresolved repressions or "blind spots." This is exactly what Stekel had failed to learn, in Freud's opinion. It echoes a critique that he had brought forward from 1908 onward, when Freud said at a meeting of the Vienna Psychoanalytic Society that Stekel "has a flair for psychogenetic complexes ... the trouble is that he relies exclusively on his inspirations, instead of submitting them to the control of conscious thinking" (quoted in *Minutes*, Vol. II, p. 10).

Contrary to Stekel, Freud never thought of Stekel's analysis as a success, nor did he consider it a training analysis. When Jung reproached Freud for having treated his pupils, in particular Stekel, like patients (Jung to Freud, 18 December 1912), Freud answered, "[It] is quite true that since Stekel, for example, discontinued his treatment with me some ten years ago, I have never said one word to him about the analysis of his own person" (Freud to Jung, 22 December 1912), suggesting that his person was not free of blind spots.

Clearly, Freud not only disapproved of Stekel's incorrigible "shallow psychology," he also abhorred the "blind spots" in his analyses as well: the fact that all of Stekel's successful cures resemble his own unsuccessful analysis. In Stekel's work, the invariable reduction of problems to a psychological condition of doubt, related to the masturbation/impotence hypothesis, seemed like an endless repetition of his aborted analysis with Freud. But Freud failed to see what Jung clearly could see: that he himself was, in a sense, to blame for that, not only because he treated his pupils like patients, thus producing "either slavish sons or impudent puppies" (Jung to Freud, 18 December 1912), but also, we argue, because of the emphasis on the element of auto-analysis in his early publications. In his attempt to appropriate Freud's methods, Stekel had imitated Freud's autobiographical style to the detail, copying his use of (covert) self-analysis in his publications. His tragedy was that this did not bring him closer to the master but, in fact, only distanced him further from him.

Rege quod est devium—"direct what is deviant," the medieval hymn "Veni Sancte Spiritus" (attributed to archbishop Stephen Langton) says–"and bend what is rigid." But as Freud no longer wished to direct and Stekel refused to bend, their ways had to part. In the next chapter we follow Freud and Stekel after their break, and examine how they would both reflect in retrospect on this episode.

Chapter 6

MARGINAL HISTORIOGRAPHY
On Stekel's Account of Things[1]

Jaap Bos

Introduction

In an unpublished letter of 30 June 1960 to Helen Walker Puner, the author of *Sigmund Freud: his life and mind* (1947), Stekel's second wife and widow, Hilda Stekel, wrote: "You were unbiased [in] what you wrote about my late husband, Dr. Wilhelm Stekel. However I feel, and so do many others who greatly enjoyed your book, that it is a pity that you wrote only about his beginnings as a psychoanalyst. His real development happened later."[1]

His real development happened later. A widow's viewpoint, certainly. But it aptly captures what Stekel himself thought about his post-1912 work, that is, his works written after his break with Freud: that they contained important improvements of psychoanalytic theory and were for that reason ignored by psychoanalysts. The catch phrase that he often employed was that he who is standing on the shoulders of a giant sees further than the giant (Freud's answer: "But a louse on the head of an astronomer does not"), in addition to the complaint that his discoveries were laughed at at first but then later adopted without proper acknowledgment. No matter how contradictory these complaints concerning the reception of his work are, they point at a problem in psychoanalytic historiography that has been widely recognized since the publication of Jones' biography of Freud (or perhaps even earlier) but not adequately dealt with: that the history of psychoanalysis is, in many ways, determined by Freud's viewpoints (see also Kuhn, 1998, pp. 1151–1152).

[1] This chapter first appeared in *Psychoanalysis and History*, 7(2005), 81–97.

History is written mainly by the victor, Paul Roazen (2001, p. 193) echoes an old saying; and if it is not the task of the historian to straighten out biases and misconceptions, it is at least his task to explore as many sides of things as possible "for the sake of promoting the cause of better understanding." Well, "better understanding" is certainly a noble goal, but we do not see how a multitude of perspectives in itself could achieve that. It seems to us that what we need is a methodology that is able to bring together all these different viewpoints: not because we want to build an unambiguous unifying framework, but to construct what we might call, following Schafer's (1992, p. 4) suggestion, a "dialogic point of view". Such a viewpoint suggests that the reading and writing of history takes place in dialogue with others; or, in other words, that it was never one voice we were listening to in the first place: the multitude of voices was already there, we just need to learn to listen to them. Such an analysis requires identification of rhetoric and argumentative as well as specific dialogic elements in historic narratives and may benefit from psychoanalysis itself, although it steers clear of psychoanalytic historiography.

In the first few chapters of this book we have attempted to explore the relationship between Freud and Stekel from this perspective, pointing out several dialogic and covert dialogic aspects in their respective works, and in particular in their autobiographic narratives. Here we would like to focus on one single dialogic dimension in that relationship: the marginalization of Stekel's historic account. For that purpose we explore Stekel's (1926) account of *his* history of the analytic movement, of which we present a complete translation in chapter 8. This account was published shortly after Freud's *An Autobiographical Study* (1925) and it was evidently meant to supplement it, or perhaps rather as its counter-account.

On the History of "On the History of the Analytical Movement"

Wilhelm Stekel's essay "Zur Geschichte der analytischen Bewegung" (On the History of the Analytical Movement) was published in 1926 in the second volume of *Fortschritte der Sexualwissenschaft und Psychoanalyse*, a short-lived periodical which he edited himself. The essay itself is dated "June 1925," which means that it was written just a few months after the publication of Freud's *An Autobiographical Study* (published in February 1925; see the editorial preface, Freud, 1925, p. 4).

In that study, Stekel is mentioned only twice: once on page 50, as one of the editors of the *Zentralblatt*, and once more on page 53, where his name appears in the list of "those who have left me, like Jung, Adler, Stekel, and a few besides [...]" (Freud, 1925). These two perfectly innocent remarks hardly compare with the harsh judgment of a good decade earlier, when Freud (1914, p. 19) said his one time co-worker had gone "totally astray" after their break, and it is therefore difficult to believe that this work of Freud was a sufficient reason for Stekel to write his extremely bitter and spiteful "On the history," in which he describes his former teacher as a terribly revengeful person who "persecuted and taunted me and tried to harm me in a narrow-minded way."[2] Yet it was.

Not only does he quote extensively from that study, he explicitly states at the beginning of his essay that he aims to "answer attacks—open as well as covert—that mean to undermine my researches," and he explains in the closing sections how he found himself forced to write his story "because the offensive account of the master is too important to just ignore and continue as if nothing happened."

"On the History of the Analytical Movement" focuses exclusively on the development of the psychoanalytic movement during the time that Stekel was part of it, and it takes as its explicit point of departure the relationship between himself and Freud. It starts with an account of how he met Freud around the turn of the century, then continues with the first (previously unrecorded) supposedly blissful years of the Vienna psychoanalytic society, when he and Freud were still on good terms with each other; it then moves on to the first congresses (Salzburg, Nuremberg), the "assumption of power" by Jung, the founding of the *Zentralblatt* and his first disagreements with Freud. Subsequently, it deals with the role of Tausk in these dissensions, then the fall of Adler and finally his own downfall, including a description of his last meeting with Freud. It ends with a general assessment of Freud's character, which is, as one might expect, far from a complimentary one.

Some of Stekel's assertions in this paper are demonstrably incorrect, such as for example his claim that he went to the 1910 Nuremberg congress unprepared and gave a presentation on symbolism impromptu which was so successful that it resulted in the creation of a research committee that was to collect dream symbols. From a letter of Freud to Ferenczi (25 Feb 1910, p. 146), we learn that Freud had the intention of asking Stekel to form such a committee months before the congress: that way he could hold him "in check" while making him serviceable to psychoanalysis at the same time. Other accounts by Stekel are perhaps less incorrect than questionable, such as his accounts of his meetings with Freud. However, a number of these accounts correspond in essence with what we know now about the history of psychoanalysis, even if they are narrated with different emphases. Given the paucity of sources Stekel could rely on at the time that he wrote his paper, we believe that "On the history" deserves not to be written off as an entirely unreliable historiography. Having said that, we want to emphasize

[2] All quotations from Stekel are from the "On the history" paper, here published as chapter 8, unless indated otherwise.

at the same time that it is not our aim to establish the truth or falsehood of Stekel's narrative. Our aim is to research it within its own scope: as a narrative with specific dialogical purposes.

Since the publication of the *Minutes of the Vienna Psychoanalytic Society* (Nunberg & Federn, 1962–1975), it has become easier to assess Stekel's account within its own terms of reference. Thus, for example, comparing Stekel's and Rank's accounts of how *The Language of Dreams* (Stekel, 1911a–c) was received by the members of the Vienna Society, it is easy to see how the former focused almost exclusively on the psychological and emotional opposition he had to face, which was, even in Rank's neutral terms, often quite strong. Of the eight speakers recorded to have spoken during the meeting of 26 April 1911 devoted to a discussion of Stekel's book, most of them praised the clinical interpretations of the dreams but severely criticized its theoretical framework as "faulty" (Sachs), "superficial" (Silberer) and "unbalanced" (Freud). Two followers of Adler (but not Adler himself) objected to that critique, which they thought was "unworthy of the rich and important content of Stekel's book" (*Minutes*, Vol. III, p. 236). When Stekel writes how "one after the other, the gentlemen pointed out to me how superficial it was," and how only Adler "who sympathized with me, spoke in an appreciative way about the effort I had made," his recollection may not be accurate in a literal sense, but neither is it wrong in essence. The same applies to Stekel's account of his closing speech at that meeting, in which he compared the reception of his book with that of a group of architects who criticize the omission of a window in the closet of a colleague's new building. Strange as the analogy may seem, and unlikely as it is that Stekel actually spoke these words, it is likely that it captures his feelings at the time.

"On the history" is thus above all a psychological self-portrait. The author does little to hide his ambiguous feelings. He is both extremely vengeful toward Freud, yet at the same time he feels the need to express his gratitude. The result is a strange mixture of overstatement and self-pity: while all the noble feelings are his (*he* can forgive and forget; *Freud* cannot), and everything that went wrong between him and Freud was surely not his fault, *he* was nevertheless victimized. Why? Because Freud could not forgive him for having made important discoveries.

"It's all a pose," Freud is reported to have said once when the conversation turned to Stekel and his wish for a reconciliation. "He plays the respectful disciple and meantime assumes the privilege of a superior. He forgives me, so to speak, for all that he has done to me" (Wortis, 1954, p. 161). Freud's witty but vicious disclaimer points to another interesting discursive feature of the historical narrative, and that is its intended readership. The essay is presented as an objective historical account whose only aim is to "serve the truth." It thus addresses nobody in particular, yet at the same time it is all too clear that it was meant to be read by Freud, and that it was designed with a specific purpose in mind: to justify himself before the master. And although nowhere does he actually address Freud directly, in several passages Stekel does come close. For example, when he refers to Freud's "autobiography," quoting the author explaining that he has had no time to return to the investigation of the "actual neuroses," Stekel retorts: "True! But what about the claim: 'Nor has this part of my work been continued by anyone else?' [...] *I have continued these studies* [...]" (emphasis added).

He desperately wanted Freud to understand him. Did Freud read Stekel's "On the history of the Analytical Movement" and understand? We have no indication that he did either. Neither Freud nor any of his followers ever mentioned it. But, interestingly, many of the claims made in this essay were already known in the psychoanalytic community, as can be gathered from the recently published *Rundbriefe* (circular letters) of the secret committee. Thus, for example, on 5 October 1920 Freud and Rank write in their circular letter that the main gist of Tannenbaum's short-lived periodical *Eros and Psyche* (edited by Silberer & Stekel) was "to reconcile official psychoanalysis with Stekel" (Wittenberger & Tögel, 1999–2004, Vol. I, p. 62), while Abraham mentions a year later how, in that same journal, Tannenbaum published a particular "Stekel inspired" erroneous assertion, namely that even though all psychoanalysts ignore the four laws of symbolism of Stekel, they nevertheless use them in their practice. "His observation contains a hidden truth: if all psychoanalysts ignore Stekel's laws, then through this *lapsus calami* [slip of the pen] Tannenbaum proves himself to be no psychoanalyst" (Wittenberger & Tögel, 1999–2004, Vol. I, p. 256).

Far from acknowledging the insights of a "dwarf standing on the shoulders of a giant," Stekel's position was never acknowledged in psychoanalysis as legitimate. Incidentally, the giant and dwarf aphorism, as Robert Merton has shown in his brilliant *On the Shoulders of Giants* (1965), goes back a long way in the history of science. From Isaac Newton to John Stuart Mill (a volume of whose work Freud translated into German), we always find it being used in anticipation of priority struggles—so in the case of Freud, when Stekel, "that unwanted disciple of Freud," tried to make "self-interested use" of it. But, as Merton aptly commented (1965, pp. 296–270), Stekel's "arrogant modesty" was "grimly crushed" by Freud's reply.

The first extant report of Stekel's use of this metaphor is in Jones's *Rundbrief* dated 11.2.1922. Jones reported to the secret committee how "Stekel's books are being translated in large numbers and he is enjoying a considerable vogue, especially in America but also here. People are attracted by his modesty! He had found the catchphrase that a dwarf on a giant's shoulders sees further than the giant" (Wittenberger & Tögel, 1999–2004, Vol. III, p. 50). To which "Vienna"—Rank and Freud—replied on 21.2.1922: "Stekel's statement should be corrected, in as much as a louse on the head of an astronomer sees no farther than the astronomer looking through his telescope" (Wittenberger & Tögel, 1999–2004, Vol. III, p. 62).

While it was again Jones (1953–1957, Vol. II, p. 136) who in his biography of Freud communicated this stinging reply to the world, it was *Freud* who had first brought the giant and dwarf trope into circulation between them when he once remarked to Stekel in private (at the time of Adler's dissension, in 1911–1912): "I have made a pygmy great, but I have overlooked a giant close at hand. A single one among the many dream symbols you have discovered is worth more than the whole 'Adlerei' put together" (quoted in Wittels, 1924, p. 225). When Stekel began using this analogy against Freud in the early 1920s, he provided it with a "theoretical context": did not Freud in his 1910 "The Antithetical Meaning of Primal Words" explain how the words 'big' (giant) and 'small' (dwarf) once meant the same? "Does that not mean that every giant is also a dwarf, and every dwarf a giant?"

Freud's louse-reply rectifies the giant mistake he had made ten years earlier calling Stekel a giant, and ended the whole play by simply forbidding the louse on his head from peeping though his telescope—at once phallic symbol and symbol of his mastery of psychoanalytic technique.

From this it follows that even though Freud may never have read Stekel's essay, he was certainly well aware of the author's intentions and was determined to put to an end all those improper priority claims. What is well known, for example, is Freud's response to Joseph Wortis, who on one occasion in the course of his analysis with Freud in 1935 passed on to him Stekel's compliment that he (Freud) was "a genius." Freud realized that it was "purposely meant for his ears" and rejected the compliment. "Calling me a genius is the latest way people have of starting their criticism of me" (Wortis, 1954, p. 142). But not only was Freud well aware of Stekel's intents; Stekel too was aware of Freud's view of him. Stekel knew, for example, that Freud had written to Tannenbaum that he, Stekel, was "a terrible man," and he also believed that Freud had said he was one of his worst enemies. This was not an exaggeration, judging not only from secondary sources (such as Wortis), but also from primary ones, such as Freud's letters to Jelliffe (Burnham, 1983, pp. 224, 252–253, 279), where we read how Freud confessed to the latter that he doubted his intentions merely because he had spotted him once in the company of Stekel.

Freud developed a complete and absolute aversion to Stekel; in dismissing his "critiques disguised as compliments," Freud not merely denied Stekel's re-admission to the society, but, in fact, he suppressed the entire account. It is significant that Stekel's essay was never reprinted anywhere, nor ever translated into English. Historians seldom found their way to it, preferring the *Autobiography of Wilhelm Stekel* (edited by Emil Gutheil) to this narrative. But while the *Autobiography* partly covers the same grounds, it was published much later (first between 1947–1949 as a series in the *American Journal of Psychotherapy*, then in 1950 in book form), and, more importantly, it was written at a time when a reconciliation with Freud was no longer an issue. The autobiography, begun on his flight from Vienna, was still not complete when Freud died, and could therefore not have the same dialogical intent (it was still unfinished when Stekel took his own life). Also the question arises as to how much the editors have altered, added or made changes to make it a coherent book. There are some noteworthy differences between the accounts in the *Autobiography* and "On the history." Therefore, for example, where the *Autobiography* (Stekel, 1950a, p. 129) describes how, during the dramatic confrontation between Freud and the Viennese rebels at the 1911 Weimar congress, "tears were streaming down [Freud's] cheeks," the same event is described in "On the history" as follows: "tears were *in* his eyes" (emphasis added). A subtle but significant difference that may be attributed to an overenthusiastic translator, but also may point to something else. The former account is quoted tongue in cheek by Gay (1988) as an example of Stekel's "melodramatic account" of the events. The latter accounts, however, published earlier and by Stekel himself, seem more in accordance with Wittels' version, which is not considered "melodramatic." On the contrary: it must be considered to have been written from Stekel's point of view, wrote Freud on 18 December 1923 to the author (Freud, 1960, pp. 350). Another subtle difference. The two sections entitled "Nervous Anxiety States" and "The growth of psychoanalysis" in the *Autobiography* (Stekel, 1950a, pp. 118–122) largely coincide both in form and content with the account in "On the history" of how his 1908 book *Anxiety Neurosis* was published (pp. 550–552 in the 1926 German publication). Freud had promised to write a Preface for it on condition that he could read the entire book beforehand; Stekel agreed and visited Freud every Sunday to receive Freud's commentary. The punch line in the story takes place when the book was finished: after much pressure, Freud delivers a cold, detached and ultimately unacceptable Preface. How deceptive and ungrateful a master Freud was! However, note that the account in the *Autobiography* is much shorter because it is stripped of all its polemics, such as a passing remark that "Freud would gladly share his minor ideas with his pupils, and not impart his more important ones to anyone" and the occasional exclamation "Regis voluntas suprema lex discipuli" [The pleasure of the king is the supreme law of the disciple]; in particular, there are no accusations, such as the claim that most analysts were financially completely dependent on Freud and therefore slavishly followed the master. Unbefitting as they may seem for an autobiography, it is precisely these polemical remarks that make the 1926 account valuable as a historic "document humain" (to use Stekel's own term).

Lastly, even though the *Autobiography* has been tidied up (either by Stekel himself, or the editor, Emil Gutheil, or both) and thereby has gained in coherence, it has at the same time lost some of its consistency. For example, the famous exclamation "I was the apostle of Freud who was my Christ" (Stekel, 1950a, p. 106) appears merely pathetic without the context of "On the history," which explains how, according to Stekel, Freud wanted a "Christ" (Jung) to lead the movement, *and* without the Jehovah parable, in which Stekel compares Freud to the furious Jehovah (*his* god) who "plunged his most beloved son into misfortune."

All this is not to say that, compared to the *Autobiography*, the 1926 "On the history" account is more "objective" in the traditional sense of the word.

One need only read the first few sentences to realize that it is anything *but* objective: "Our memory is partial. Freud has proven this and presented numerous examples from which this conclusion follows. When I try to present an account of the history of the analytical movement, I can assert that such a 'partial functioning of the memory' has not strongly developed in me. I have no loss of memory." Surely a peculiar thing to say for someone who has studied psychoanalysis for so long. But that is exactly the point here: had it been an "objective," or even remotely neutral account of psychoanalysis, there would not have been any reason to bother reading it: it does not record anything that we do not already know by now (although in 1926 it was probably "revealing" in a sense).

This, then, is to say that "On the history of the Analytical Movement" should be read as a *dialogue*. And that is what we explore in this chapter.

More Freudian than Freud but not a Freud

Stekel enjoyed considerable popularity, both in the German speaking world as well as in the United States. He had a school of his own and liked to boast about his short-term therapies, which, unlike the long-term Freudian analyses, were always successful, so he claimed. He was a heretic, and proud of it too because it was Freud, not he, who had gone "totally astray," particularly with respect to lay analysis: he maintained that until approximately 1910 "Freud was still strongly against lay analysis." To support this claim, Stekel mentions the case of Viktor Tausk, a lawyer who "burned with ambition" and devoted his life to psychoanalysis but "did not want to analyse as a lay analyst" and therefore "began to study medicine and eventually became a practicing doctor." "Tausk" was the catalyst for the process that led eventually to the rift between Freud and Stekel. But Tausk was only the "displacement" of a hidden figure: "Tausk was a mere battering-ram, used by Freud to take stronghold Stekel."

The reasons that Freud would tolerate lay analysis were, according to Stekel, perfectly opportunistic: firstly Freud needed to compensate for the loss of "important medical men" such as Adler, Jung, Bleuler and himself; secondly, he wanted to pass on his legacy to those who had remained "faithful" and happened to be laymen, and lastly perhaps also to protect his daughter Anna. We will return to the problem of lay analysis below; suffice it here to note that to Stekel none of these arguments stood up against his "scientific critique": "Freud knows very well that he would not have arrived at his analytical discoveries without medical knowledge."

Stekel therefore had reasons for regarding himself as more Freudian than Freud: "I now represent Freud much better than he does," and he criticized Adler and his followers for still being Freudians despite themselves: "It would be a tempting piece of work to analyse some of our individual psychologists and to prove through them the existence of the very Freudian complexes which they have banished from their books," wrote Stekel (1923a,b, p. 527) in a little essay not long before the "On the history" account.

But despite his success, or perhaps *because* of it, he felt deeply hurt about the fact that Freud refused to acknowledge his position as an analyst. Freud would go no further than to admit that he had "remained loyal" and "of use" to psychoanalysis, adding in the same breath, however, that he had "done it also great harm." That was in the only letter of Freud to Stekel that has been published until now. Significantly, that letter was written on 14 January 1924, half a year before Freud began writing his *Autobiographical Study*, and evidently in reply to one of Stekel's attempts to reconcile with the master. "I broke with you after you had deceived me in a most heinous manner (you never mentioned this occasion—*Zentralblatt*—in your letters)," Freud wrote. What is interesting about the sentence in parentheses is that it could be read not just as an accusation (You betrayed me, don't you remember?), but also as a challenge, or perhaps even an invitation: If you want to reconcile, you must explain to me why you behaved so badly on this occasion. Stekels reply of 22 January 1924 evades the question: "the past is not to [be] discussed."

Could not Stekel's entire historic narrative be considered a response to that challenge, triggered by Freud's autobiographical account, in the hope not only of re-establishing relations, but also of forcing Freud to acknowledge *his* errors? If so, it was certainly an unrealistic expectation, given Freud's earlier public statement about his one-time collaborator having gone "totally astray." But it is understandable given the nature of their relationship. In our chapter on Stekel as a silent antipode, we characterized the Freud–Stekel relationship as driven by a "dialectic of antagonism." The implication of that idea is that, although Stekel frequently contradicted Freud, both benefited from this conflict of viewpoints: Stekel because he found acknowledgment in Freud's contradiction; Freud because he could use Stekel's counter position in his didactic discourse. A clear example of such a use we found in Freud's *Gradiva* study, which was hitherto believed to have been written in order to do Jung a favor. In fact, we argued, it was written as a tutorial analysis, with Stekel serving as the rebellious student whose position is used to teach to others the appropriate methods.

This antagonistic relationship ceased functioning sometime after 1911–1912, not so much because Freud suddenly felt fed up with his one-time counteractive co-worker or because he could not forgive him for the "*Zentralblatt* treachery" (although that may have been the straw that broke the camel's back), but because relations within the analytic movement began to change as a result of institutionalisation processes, as a consequence of which dimensions of Freud's authority changed as well. The key aspect therein seems to have been the fact that Freud no longer exercised his authority in face-to-face situations but had to rely on written communication more and more heavily. As a consequence, Stekel's antipodial position was no longer serviceable to Freud. After that, their relationship disintegrated very rapidly, leaving both men with more longer-lasting hostile feelings for each other than would seem functional.

It is not necessary for this hypothesis to be plausible to assume that either one of the parties was aware of this antagonistic dialectic in their relationship. In fact, from a psychoanalytic point of view it may perhaps be considered a necessary condition for them *not* to be aware of it, of it being a "blind spot" of Freud's, so to speak. But that is not to say that Freud and Stekel were not aware of the consequences of the *end* of their antagonistic relationship. Embittered, Stekel wrote how Freud had "no use" for him anymore; Freud, equally embittered, wrote to him in his letter of 13 January 1924 how he had at one time been "of use" to psychoanalysis, but now collaboration was "no longer possible."

Emphasis on functionality in this antagonistic relationship should not, in our opinion, be confused with an interest in individual morality or personal character traits. We are not dealing with an *intra*-personal dimension here, but with a particular twin-faceted *interpersonal* dimension: marginalization. When Roazen (1986, p. 48) wrote that "for a man really to be like Freud meant finally for him to be original. Yet originality ended his usefulness to Freud," he illuminated in that attractive paradox one side of this dynamic, emphasizing the role of the oppressor. The other side of this dynamic, that of the oppressed, remains out of sight. It is this side that has found an equally attractive paradox in Stekel's assertion that he was more Freudian than Freud—but not a Freud. He did not want to put himself on a par with Freud, and never wanted to claim that he himself was a genius, yet in all his writings he attempted to prove that he was at least his equal. "I received a great deal from Freud. But I gave at least as much, if not more. We could call it quits."

Bourdieu (1992, p. 121) emphasizes how the ruler succeeds in imposing his views on the less powerful, he speaks of a "specific kind of communication" between them, where the one "signifies to [the other] what his identity is, but in a way that both expresses it to him and imposes it on him by expressing it in front of everyone [...] and thus informing him in an authoritative manner of what he is and must be." Here we want to emphasize that it only rarely comes to this point, and that, in general, marginalization is a process that takes place with the full consent of the marginalized; in fact, he often actively contributes to it—that is, to his own marginalization—not out of sheer masochistic pleasure, or because he is a victim of a Foucauldian disciplining machine, but knowingly and willingly, because the marginal position is a position in its own right. So it is in the case of Stekel, whose marginalization was a beneficial, not a destructive, factor in the construction of his position in the psychoanalytic community. To refer to Stekel's deviating works as proof of Freud's non-authoritarian style of leadership, since the master left his followers "free to indulge in errors of their own choice"

(Eissler, 1965, p. 386), is to mistake the co-construction of marginalization by the one who is marginalized for indulgence on the side of the one who marginalizes.

With the co-construction of marginality as our main focus, let us now return to the question posed at the beginning of this paper: how to discern all the various voices that constitute historic narratives? More precisely: how to differentiate between the voice of marginal and the marginalized in a historic account?

Deep marginality

In this analysis of Stekel's marginal voice within psychoanalytic historiography we have focused on the identification of various more or less explicit references to, links to and comments on Freud's texts (both private and public ones) by Stekel, or, to use a more modernistic term: on the exploration of "manifest intertextuality" (Fairclough, 1992). The conclusion it has led us to so far is that, from a dialogic point of view, Stekel's own work contributed to his marginalization. In this section we shall elaborate on that idea and explore the links that are less explicit or visible; that is, the analysis of "deep marginality."

The first "deep links" to be explored are those that Stekel himself addressed. Of these, his comments on the long passage from Freud's *An Autobiographical Study*, where Freud explains how he came to differentiate between "actual neuroses" and "psychoneurosis" and does not mention Stekel anywhere, are, to begin with, the most interesting. It was this autobiographical study of Freud that triggered Stekel to write his account of things; however, as has already been noted above, he is actually mentioned only twice in this work, and even then only in neutral terms. Therefore, the question is: what exactly bothered him?

The sting in the quote from Freud's *Autobiographical Study* is located in its tail, the last sentence, where Freud bluntly asserts that "this part of my work has not been continued by anyone else." Stekel had always maintained that all of his psychoanalytic publications, most particularly his book on anxiety neurosis, were a continuation of precisely that part of Freud's work. "Has he really forgotten the book I wrote," Stekel asks in indignation. The book was *Nervöse Angstzustände und ihre Behandlung*, published in 1908 and "extremely successful" (not an exaggeration by Stekel). For that book, Freud had written a preface (Freud, 1908d) in which he however limited his own responsibility to two words: the term "anxiety hysteria." An unsolicited but "royal" gift that was reclaimed immediately by the master, as Stekel bitterly observed. Well, he could have it back: after 1912, from the third edition onwards, Stekel no longer used Freud's preface, nor his gift: "His viewpoints have been separated from mine now." An interesting choice of words, suggesting it was Freud who had made the separation.

But it is not merely in the conclusion of that quote, in the tacit denial of Stekel's contribution to the problem of anxiety hysteria, where Freud's former follower felt betrayed. Note how, in his discussion of neurasthenia, Freud mentioned how one of its causal factors is "excessive masturbation." This was of course another major antagonistic element in Freud's and Stekel's working relationship. There is reason to believe that it already played a role in Stekel's motivation to seek counselling from Freud (see chapter 3). In his historic accounts Stekel euphemistically called these motives "confusions" (or *Wirrnisse*, in the German original), suggesting his problems were of a semantic origin (but see chapter 5 for a more extensive account of Stekel's therapy). At any rate, whether or not Stekel's insistence on the harmlessness of masturbation was induced by his own "confusions" or not, it became a central dogma in his analyses, and he related half of his cases or more to this problem. As if he wanted to aggravate Freud by showing one patient after another who became ill because he had to *repress* the urge to masturbate. Indeed, an unforgivable act, so to speak.

The next link to be found, also in a discussion of a quote from *An Autobiographical Study*, is the hypothesis that lay analysis is harmless. Freud defended the weaker hypothesis, while Stekel found most analysts on his side (orthodox and otherwise, a factor he failed to mention). Why did Freud defend lay analysis? Because "important doctors," such as Stekel himself, had left the movement, and only "unimportant laymen," such as Rank, Reik and others, remained loyal. Perhaps also to defend his daughter, Stekel said somewhat shabbily. But also, we will add, because "a doctor who has not been through special training is, in spite of his diploma, a layman in analysis," as Freud (1925, p. 70) wrote. How much imagination do we need to realize that Stekel once again recognized himself in these words? Stekel had not been in analysis with Freud. He had had therapy, and even that only briefly, not "that special training"; so technically, that would make him a layman in Freud's eyes. "You should mutually analyse one another," Ferenczi suggested to Adler and Stekel in 1910, when they travelled home together from the Nuremberg congress, "to overcome your brother-complex" (Ferenczi to Freud, 5 April 1910). In Stekel's opinion that would not have done. He and *Freud* should have mutually analysed each other. After all: "It is Freud's tragedy that he has never been analysed himself. He needed it as much as many of his students."

This claim puts us on the track of another intertextual link which is related to the quotation from C.F. Meyer with which Stekel ends his paper: "I am not a book that's filled with clever fiction. I am a human being, in all my contradiction." Stekel used these lines in several places in his works, for example, in 1935, in his *Fortschritte und Technik der Traumdeutung* (oddly translated as *The Interpretation of Dreams*), when he stressed how important it was "to mental equilibrium, to the balance of the mind, that (as the law of bipolarity prescribes) in the mind there should be contrasts, holding counterpoise in the scales" (Stekel, 1943, p. 139). This, of course, is a reformulation of Freud's argument, who, in the case history of Little Hans (Freud, 1909, p. 112), wrote that "we ourselves, too, must guard against making a difficulty of such a contradiction. The emotional life of man is in general made up of pairs of contraries such as these," and then, in a footnote, gave the two lines from C.F. Meyer without further commentary.

However, instead of stopping short of just paraphrasing Freud, we find that right after that quotation from Meyer, Stekel (1943, p. 139) continues with the following argument: "This introduces us to another and very important question. Is auto-analysis possible?" The answer is: no, because "Freud, in his *The Interpretation of Dreams,* overlooks important complexes where his own dreams are concerned. My gratitude to this great master, from whom I myself learned the art of interpretation, has always restrained me from the attempt to analyze his dreams [...]."

The C.F. Meyer quotation thus represents Freud in at least two ways, and thus does two things at the same time. First it allows Stekel to pose as Freud's equal: his law of bipolarity is at least as good as Freud's. Secondly, it allows him to surpass Freud: he understands the master's blind spots better that Freud does (see the chapter on the art of imitation for further analysis of this idea).

There is a final "deep link" of intertextuality to be explored; this link, however, is neither addressed directly by Stekel, nor implied in Freud's texts, yet it is persistently present in both accounts. It is the question of why Freud did not say something publicly about his break with Stekel. All Freud would say was that he found it "difficult to publish an account of." In his letter to Stekel he certainly put the blame of the addressee's "personal qualities—usually described as character and behavior," not their "scientific differences" (Freud to Stekel, 13 Jan 1924). But that was unacceptable to Stekel, not, because of the insult that followed ("Nature has endowed you with an unusual degree of self-complacency"), but because Freud chose to ignore that antagonistic relationship with Stekel completely—as if there had never been one. Thus Stekel replied on 22 Jan. 1924: "The same incident in the minds of two people can look very different" – as if they were still in some kind of dialogue with each other.

"I have elsewhere always been ready to acknowledge what I owe to other workers; but in this instance I feel burdened by no such debt of gratitude," Stekel quotes Freud's Preface to *The Ego and the Id* (1923b, p. 12) only to continue his exposition with a recollection of how Freud once reproached him: "You thought you could forget me!" It is easy to see how in that quotation from the preface to *The Ego and the Id,* where Freud explains why he will not refer to "the theories put forward by non-analysts or by former analysts on their retreat from analysis," Stekel both feared *and* desired to be identified as a "former analyst," so that he could retort with Nietzsche: "You serve your master wrongly if you don't transcend him."

The entire undertaking of his historic essay was thus to receive recognition, not for the merit of having served Freud but for the virtue of having contradicted him. What Stekel demanded from Freud was a continuation of the antagonistic dialectic that had driven their relationship. Contradiction: acknowledgement: contradiction: acknowledgement—the only true "objectivity" he could demand from Freud.

The "deeper intertextual links," as we described them here, are those that refer to what is missing from discourse and at the same time presumed to be at work. In that sense we must consider the references to anxiety neurosis, masturbation, lay analysis and objectivity as *agents provocateurs* in Stekel's historic account: shadows of the angry demon that disturbed their happy marriage.

Conclusions

To conclude, let us briefly sum up the findings, advantages and disadvantages of this little study. First we argue in this chapter that the marginalization of Stekel's historic account is a "co-construction" of both the authoritative source and the "victim" of marginalization; second we found that since marginalization is a dialogic rather than an inscriptive act, the study of it should rely on the identification of specific discursive processes, that is, intertextual links.

The advantage of a dialogic approach in contrast with other approaches is, we believe, that we are allowed to identify multiple voices within historic discourses, instead of identifying multiple but disconnected discourses voiced by various authors. Perhaps this method also allows researchers to use multiple explanations more readily, including psychoanalysis itself.

In order to investigate a subject within a field, one should, we believe, try to use criteria that lie outside that domain—which is not to say, however, that the researcher can be ignorant about its content. But knowledge in the field should not be taken as a standard to investigate that same body of knowledge. In this book, we approach the history of psychoanalysis as a *practice* in which actors take part, and which is shaped by these actions and finally becomes a framework within which the actors operate on the world. We seek to uncover the logic in or behind these practices by focussing on "discourse." By applying a technique that has been used in ethnomethodology and sociology to study ways in which ideology and individual action cooperate to produce stable patterns of interaction in practices, we explicitly try to avoid separating discourse from body. Discourses are embodied intentions, not merely the expression of an intention.They convey a message, but also do more than that: they invoke in the reader expectations, anger, laughter, etc. It is more than linguistic usage to ascribe intentions to texts: they actually have, as John Austin called it, "performative power." They can make us do things.

Texts (as a specific form of discourse) are thus more than a simple expression of actor's intentions—they are the constructions people reside in, to borrow the words from a romantic Dutch poet.[3] Similarly historic narratives are more than reconstructions of events: they are narratives told and lived at the same time. One might say that the entire psychoanalytic enterprise would not be possible without the last part of this claim. If we did not live but only told our narratives, we would neither be frustrated if we could not live up to our own and other people's expectations, nor would we want to change our perspective if we felt hampered by our own history.

But this clearly also points to some of its disadvantages. We will restrict ourselves to pointing out an important one, which is that a dialogic approach could easily be mistaken for psychological *"hineininterpretieren"* (ascribing motives, intentions, etc. to actors as the driving principles behind their actions). However, we would not like to invite a psychoanalysis of the analyst ("the autobiographical trap"). We wish to avoid the endless speculation about hidden motives, secret intentions, dark drives, etc. that some historians are fond of but in our opinion lead us nowhere. The ineradicable Freud-Minna Bernays "affair" is a good example of such an over-analysed speculation. Assuming the liaison were true, the more interesting question would be: what interest does that affair have to psychoanalysis? What new insights or understandings could it reveal? We find these questions never answered other then in terms of ever more speculation. In that respect, the history of psychoanalysis sometimes resembles a *perpetuum mobile* of interpretations: each interpretation brings in motion another one, and they all go back to Freud. To maintain with some historians, that *everything* Freud wrote was self-analytical is reducing all the various genres Freud used to only one, and this merely seems to serve as a licence to probe each and every word of Freud, and find hidden autobiographical meanings behind all of them.

Therefore, to identify in Stekel's *discourse* a desire to be acknowledged by Freud does not equate with interpreting a psychological motive, nor does it equate with psychoanalysing the psychoanalyst. The advantage of using psychoanalytic tools is limited to a *discursive* level only; that is, to an analysis of conflicting themes, implicit premise or explicit assumption within discourse at an intertextual level. The marginalization of Stekel is not, in my opinion, a problem of personalities. Varying on Michael Billig's (1999) formulation, we might say that a marginal voice is not a voice repressed from consciousness, it is one of the many latent voices we can find in public discourse which have been prevented from articulation.

[3] J.J. Slauerhoff: "Only in my poems can I reside."

Chapter 7

EVIDENCE OF ESTRANGEMENT

Notes on the Correspondence of Wilhelm Stekel to Sigmund Freud

Jaap Bos and Paul Roazen

Introduction

The fracture between Freud and Stekel belongs to one of those "fateful moments" in early psychoanalytic history about which there is a fair amount of information. The issue receives ample attention in virtually all Freud's letters from this period (see the Freud–Jung and Freud–Ferenczi correspondence in particular). Moreover, scattered throughout the collected works occasional remarks on this event come to the surface (Freud, 1914; 1925), while historians of science were able to fill in some of the missing information (Jones, 1953–1957; Roazen, 1975; Gay, 1988). There is, furthermore, a fair amount of coherence between these sources. Many agree on the opinion that this was not a major event in Freud's life and that, in 1912, his patience with Stekel had simply run out. It was not scientific differences that led Freud to decide to break with Stekel but only personal motives: the "character and behavior" of the other, as Freud wrote to Stekel on 13 Jan. 1924. Understandably perhaps, Stekel strongly objected to this conception, but he was unable to convince historians of his viewpoints and was thus destined to become psychoanalysis' marginal. It does not alter the fact that the *dynamics* that led to this fracture have remained largely obscure.

In this chapter we aim to recover these dynamics by exploring Stekel's side of the story more fully. By thereby completing the picture, it is not our intention to rewrite history. Nor do we wish to attach more importance to the rupture than has already been given, and neither do we aim to cast doubt on Freud's motives (or Stekel's, for that matter). Our aim is to provide an insight into the *interactional logic* that led to Stekel's drifting apart from psychoanalysis. Ultimately, this may help us understand the growth and transformation of psychoanalytic discourse in its early history.

To do so, we discuss the hitherto unpublished Freud–Stekel correspondence conserved in the Freud Archives of the Library of Congress (Washington), as well as four related items (including one letter of Stekel to Jung). The correspondence is clearly incomplete, in as much as only five letters of Freud have been preserved, and that all of Stekel's letters prior to 1910 and a number from after 1912 are missing. Therefore, while the correspondence formally covers the 1904–1938 period, the bulk of the letters date from the crucial years 1910–1912. In other words, they record the main period of estrangement between the two men. Chapter 9 presents the full correspondence for the first time in its incomplete form as it is preserved in the archives.

While the reasons why Freud kept this part of the correspondence (and presumably destroyed the rest) are unknown, the fact that they cover their break implies, we believe, that he might have kept them for reasons similar to organizations when they keep a "paper trail" of an unwanted employee: as a collection of evidence in a case against him. Consequently, we read these letters as documentation of a particular historic relationship on the one hand, and as the exact opposite of that on the other: as proof of its deterioration.

By way of introduction to the correspondence, we present a brief overview of the historical background against which the letters are set and discuss the dialectics of this estrangement.

Historical Background

The earliest items from the Stekel–Freud correspondence comprise two short notes from Freud dated 1904 and 1905 respectively, interesting in themselves from an "evolutionary" point of view: as missing links that bridge the gap between the end of the Fliessian period (1904) and the beginning of the orchestrated correspondence with his followers (1907). Thus Freud's 1904 note plainly acknowledges Stekel's role in the foundation of the "psychological society," a fact that he would come to veil in later years (Freud, 1925, p. 43).

There is a five year gap between Freud's notes and Stekel's first preserved letter of 28 February 1910. During these years, the local, Vienna-based informal group of followers had developed into a movement with followers in several European countries and America who had gathered together for the first time in 1908 in Salzburg. With the second congress scheduled to take place in Nuremburg in March 1910, the movement was about to develop into a formal and professional organization, with its own institutions, written and unwritten rules, etc. It is at this point that the Stekel–Freud letters commence.

To Freud, the Nuremburg congress clearly marked the end of his "splendid isolation." Correspondingly, his opening address (Freud 1910d) was a daring, sweeping, optimistic speech about all that psychoanalysis had achieved so far, and everything that remained to be expected. Freud was at the peak of his powers. He had opened up his own universe and was now ready to conquer the world.

To Stekel, on the other hand, Nuremburg marked the beginning of the end. In the run-up to the congress, his position in Vienna was damaged when he was savagely attacked after a presentation he had held on 19 January 1910 on "the psychology of doubt." Freud thought it "theoretically weak and blurred"; Sadger reduced the presentation to a mere "juggling with words" and Binswanger, present as a guest, found "nothing new" in it, but saw instead too many generalizations (*Minutes*, Vol. II, pp. 400–401; Freud and Binswanger, 1992, pp. 259–260). It is clear that it was Binswanger whom Stekel referred to when he complained to Freud how a guest had made fun of him and treated him as a "sampler" (*Kompilator*). In the conclusion of the discussions, Stekel tried to keep up appearances but to Freud he confessed that he was "shocked" and could not bring himself to attend the next meeting.

Stekel later claimed that Freud vetoed publication of his paper on doubt in the *Zentralblatt* and that he had to publish it elsewhere (see chapter 8). In his letter of 28 Feb. 1910 to Freud, however, he is contemplating letting it "mature" and then reworking it into a book-length publication. In the end he did both: he first published it elsewhere (Stekel, 1912a), and then also included it in the final volume of his "Disorders of the Instincts and the Emotions" series, on *Compulsion and Doubt* published much later (Stekel, 1962a), which indicates, perhaps, the importance he attached to this work. Indeed, in this paper on doubt, for the first time Stekel develops the idea of bipolarity, which he would come to see as one of his key discoveries: how every affect always carries its own negation in it. Freud remained unconvinced and preferred Bleuler's term "ambivalence" for more or less the same phenomenon (today in psychiatry the term "bipolarity" is used to denote manic depression; there seems to be no reference to Stekel in that literature).

At Nuremberg, Stekel proposed the founding of a committee to collect contributions on symbolism. The idea came from Jones, who had written to Freud on 12 February 1910: "Do you not think the time is ripe to apply a suggestion you made in the *Traumdeutung*, namely to make a collection of typical dreams? Why not establish a central bureau at Jung's to which short accounts of analyses could be sent by different workers?"

(Jones to Freud, 12 February 1910). Freud must have communicated this to Jung (the letter is missing), as the latter responded: "The *group research on symbolism* is an excellent idea. I have long been thinking along those lines" (Freud to Jung, 2 March 1910, emphasis in the original). Shortly before the congress, however, Freud wrote to Jones (who would be unable to attend) that he was going to accept his symbolism proposal, but that he would "instigate Stekel to put it before the Congress. No doubt, it will be executed in the way you intended [...]. It is a good way to get him under control."

Typically, in later years, Stekel appropriated the idea as his own, when he claimed that he went to Nuremberg unprepared and gave an improvised presentation that resulted in the creation of this committee (see chapter 8). At any rate, a number of papers found their way to him. He subsequently published them in the newly founded *Zentralblatt für Psychoanalyse*, in the "Communications" (*Mitteilungen*) section. Among them were two short contributions by Freud, apparently written immediately after the congress (Freud, 1910a, 1910f). Freud would adopt several of these "communications" as illustrations in subsequent editions of his *The Psychopathology of Everyday Life* (Freud 1901; see Bos 2004 for a further analysis of Freud's policy of creating a unified body of knowledge from these illustrations in his works).

However, Stekel took the Jones–Freud suggestion much more seriously and soon began writing a large volume on symbolism in dreams, entitled *The Language of Dreams* (Stekel, 1911c), which appeared in early 1911 at a very precarious moment, namely when Adler's position at the Vienna society was perceived as untenable. The critique would also affect Stekel's position (as we shall see in the next section).

The pretensions of *The Language of Dreams* went far beyond what Jones had in mind and the book was criticized by many, including Freud, who wrote to Jung that it was "theoretically weak." But as an illustration of his own ideas he believed the book to be valuable, and thus added to Jung that he thought it was "rich in skilful analyses and likely to make an impression" (Freud to Jung, 14 January 1908). Strikingly, there is no mention of its development in the correspondence with Stekel, and the book only surfaces in Stekel's letters after publication, implying that Stekel refrained from seeking Freud's advice. Indeed, in his 1926 historic account Stekel wrote that he did not even want Freud to read the proofs since he had had such negative experience with his previous book (on anxiety neurosis).

The Language of Dreams turned out to be a huge success, as friends and foes alike were forced to admit. This literary success was countered by a deception which was at least as considerable. When Ferenczi proposed in Nuremberg that psychoanalysis institutionalise itself as a formal organization, the Viennese followers in general and Stekel in particular were bypassed entirely in favor of Jung, the proposed president for life of the association, who was to have "unheard of powers of censor" (Jones, 1953–1957, Vol. II, p. 96). This aroused so much animosity that discussions had to be postponed to the next day. In the evening, a secret meeting took place in Stekel's hotel room. Freud got wind of it and gave a dramatic speech, claiming, in the words of Stekel, that his "very existence was at stake" (see chapter 8). The next day Ferenczi's proposals were mitigated somewhat and then accepted. Jung became president.

In compensation for their "lost hegemony," Stekel and Adler were granted publication of a psychoanalytic journal: the *Zentralblatt für psychoanalyse*. This was to be the first regular journal accessible to a larger public that carried psychoanalytic "propaganda." It was first published in October 1910.

Evidently, much of the correspondence between Freud and Stekel is concerned with editorial matters and negotiations with publishers and printers. It is also quite clear that the involvement of Freud, presented as the journal's *Herausgeber* (editor in chief or director), was initially relatively small, and that he feared that it would "undermine his institutions" and lead to "endless polemics" (Stekel to Freud, 26 April 1910). From the Freud–Jung correspondence (Freud to Jung, 17 February 1911) we learn that Adler and Stekel took turns at editing, but it was Stekel alone who identified with the journal, and who wrote by far the largest number of reviews, "communications" and "miscellaneous" for it, which, in 1912, elicited the following remark from Freud: "It is touching of you, and a disgrace for contemporary mankind, that you wrote almost all the reviews ... " (Freud to Stekel, 11 March 1912).

All this work, however, could not satisfy Freud, who would criticize the "messy work" of the editors, and complained to Jung about the fact that the journal now had too much Adler in it, now too much Stekel (Freud to Jung, 17 Feb. 1911). It is true that while the issues were prepared well in advance, they nevertheless often did not turn out as Stekel had imagined, and one must presume that Freud interfered with editorial matters more frequently from the second volume onwards. In autumn 1911, just before the end of its first year, Adler resigned as co-editor of the *Zentralblatt*, leaving Stekel behind as its sole editor who jealously kept an eye on the competition. Thus he attempted to dissuade Freud from publishing a rival journal, *Imago*, which he feared could only harm his journal. When *Imago* was nonetheless first published in 1912 he hoped he could at least screen off his own field of interest, the medical and therapeutic side of psychoanalysis, and let *Imago* cover the literary publications. His pleas were ignored, and Stekel had to witness how *Imago* also ran medical papers.

When in late 1912 Stekel came into conflict with Freud directly and could no longer be tolerated as a member of the psychoanalytic society, Freud urged all his followers to withdraw their support from the *Zentralblatt* and founded a new journal, the *Zeitschrift für ärztliche Psychoanalyse*, which was essentially a continuation of the *Zentralblatt* but without Stekel. Stekel continued publishing his journal alone for another two years, bringing psychoanalytic papers and reviews of books of Freud and his followers as if nothing had ever happened. When war broke out the journal stopped publication (see Appendix at the end of the letters), which effectively ended Stekel's last tie to Freud. Several attempts to re-establish relationships with Freud met with a decided "no." They probably never met again after 1912.

The Fracture Between Freud and Stekel

When we deal with the fracture between Stekel and Freud, we must first of all observe that there was not one but several. The earliest recorded rupture in their relationship dates back to the very first letter that is preserved in this correspondence, when Stekel apparently, for reasons unknown, resigned from the society he himself had founded and Freud was asked to persuade him to return to it—according to Freud's 1904 note at least.

The second break (that we are aware of) occurred early in 1911, when after extended debates in January and February of that year, Adler's viewpoints were declared "incompatible" with Freud's. Adler was forced to resign as chairman of the Vienna society and Stekel followed him (*Minutes*, Vol. III, p. 177). Late February to early March, Freud announced to several of his followers that both Adler and Stekel had resigned (Freud to Pfister, 26 February 1911; Freud to Jung, 3 March 1911; Freud to Ferenczi, 12 March 1911). In retrospect, Stekel wrote that his first response was to join Adler but that he actually refrained from doing so because he felt financially dependent on Freud, and also found it difficult to give up the *Zentralblatt*.

And indeed, what followed in mid March was, it seems, a period of negotiation or arbitration, in which editorship of the *Zentralblatt*, still in its first year, presumably played a key role (see Handlbauer 1990, for a detailed discussion of the Adler–Freud controversy). On 8 March 1911, Jung asked Freud: "What will become of the *Zentralblatt* if the wind blows from that quarter? Will you replace the editors?" Freud replied on 14 March: "The palace revolution in Vienna has little effect on the *Zentralblatt*. Naturally I am only waiting for an occasion to throw them both out, but they know it and are being very cautious and conciliatory, so there is nothing I can do for the present."

A private dinner between Freud, Adler and Jekels (the latter presumably acting as mediator) did not result in a settlement of the conflict, but neither did it yet result in a clear-cut break. Adler continued to attend the society until 24 May 1911, although he no longer actively participated in the discussion (see Jones, 1953–1957, Vol. II, pp. 132–133; *Minutes*, Vol. III). By mid June Freud reported the "removal" of Adler (his official withdrawal from the society took place in October).

Stekel, however, managed to reconcile with Freud. At the beginning of May, Freud mentioned to Jung that he "has decided to put up with Stekel again." The price Stekel had to pay for his settlement was a distancing from Adler with whom he had always been on good terms. In return, Freud withdrew a critical review he had written of Stekel's recently published book on dream symbolism (he read the review, however, at the meeting of the Vienna society on 26 April 1911). The next day Freud reported to Jung about the incident thus: "I read the review in question aloud; he [Stekel] reacted as if he didn't feel spat upon but had only felt a few drops of rain. So it passed off quite well" (Freud to Jung, 27 April 1911).

That summer, Stekel wrote to Freud to say how delighted he was to "be able to love his teacher again," and, while on holiday in Ischl, he reported somewhat obsequiously on Adler's reservedness towards him ("even formal salutations are missing ... "). He was planning a special issue of the *Zentralblatt* in "tribute to Freud," apparently to celebrate the third psychoanalytic congress but obviously intended as a peace offering of a kind. The issue appeared as the last edition of the first volume, carrying a good dozen short papers (and one long one) by loyal Freudians, among which, interestingly, Rudolf Reitler's "Critical remarks concerning Dr. Adler's 'male protest' doctrine," which had given the impetus to the Freud-Adler alienation. Stekel was rewarded for his loyalty with a valuable plate; a note from Freud (missing from the correspondence) says "I don't know what could ever tear us apart ... " (quoted in Stekel's 'On the history' chapter). In an interesting letter, from a psychoanalytic point of view, Stekel wrote to Freud in the summer of 1911: "I am recovering well. *You would not recognize me*" (emphasis added).

As the new semester started in October, Stekel reported to Freud that he got word from Rosenstein that Adler's followers were planning a collective withdrawal from the Vienna society: "It would be a tactical error if we were to allow them such a nice dramatic retreat". Thus, at the society's first meeting, on 11 October 1911, and in anticipation of this move, Freud ruled out dual membership: "those who also belong to Adler's circle [are confronted] with the fact that its activities bear the character of hostile competition" (*Minutes*, Vol. III, p. 281). Members had to choose between the two societies. After prolonged discussion, in which Stekel somewhat hypocritically attempted to play the role of mediator (Nunberg & Federn, p. 282), the Adlerians resigned from the society.

Unfortunately for Stekel, his reconciliation with Freud and his break with Adler did not improve his position in the Vienna society. That summer a series of discussions on the problem of masturbation, a subject to which Stekel attached great importance, took place at the Vienna society in the winter of 1911–1912. While Stekel was convinced that suppression of the urge to masturbate causes feelings of guilt, and that feelings of guilt cause "nervousness" (anxiety neurosis), Freud claimed, not, as Stekel believed, exactly the opposite (that masturbation itself is a cause in nervous anxiety), but rather that masturbation is "not anything ultimate," that it is 'not a real "agent'" (Freud, 1912b, p. 251).

The contributions to the debate were collected in the spring of 1912 and published later that summer. Freud's concluding remarks deal at large with Stekel's dissenting views, which were dismissed as "non-psychoanalytic" (Freud, 1912b, p. 248). While ignoring the criticism, Stekel congratulated himself in a letter to Freud of 17 Aug. 1912 with the amount of attention the latter devoted to him and presented his own views as a "necessary counterbalance" against "the various voices which we have heard in the discussions" (Stekel to Freud, 10 April 1912).

In early spring 1912, it became clearer that Stekel now found himself in pretty much the same position as Adler had been a year earlier: his viewpoints were in open contradiction to those of Freud, and by insisting on them as he did, he had effectively positioned himself outside psychoanalytic discourse. Freud's followers no longer wanted to associate with him and even avoided his company. Stekel was becoming an "outsider" and consequently dependent on Freud's forbearance—a fact he himself emphasized in his 1926 historical narrative when he wrote how he had become once more "dependent" on Freud (see chapter 8). To Freud he complained on 1 June 1912: "I do not feel surrounded by friends. It has actually become terribly uncomfortable for me although I have certainly done everything I could to make friends." At the end of June he went to Ischl to spend his holidays.

Once there, he resorted to a hazardous strategy to regain Freud's trust. When he had been in Ischl the previous summer, Stekel had run into Freud's mother (Amalia) and Freud's sister (presumably Adolphine, who often took care of her mother), and discovered that they lived opposite him ("think of the coincidence ... "). He must have found out that "mother" spent her holidays there every year. As he left on the 25 June 1912 for Ischl ("same address as last spring ... "), he wasted no time. Four days later a letter arrived:

> Dear Professor!
> Yesterday I had a consultation with Dr. Steinsberg and examined your mother. It is a bad case, but definitely progressing, which, however, the terrified patient would never admit.

The nature of Freud's 77-year old mother's physical problems remains unclear. 1912 saw a particularly bad summer. It rained constantly and "Frau mama Freud" was confined to her holiday retreat. In subsequent letters there is mention of pains in general, "bad nights," a swelling of the hands, but also of complaints about a "lack of sunshine" and too few tarok games. One cannot avoid the impression that she may have suffered from no more than a case of boredom. At any rate, Stekel immediately took over the "treatment" of Freud's mother and recommended all kinds of measures, such as the application of moist heat packs, inhalation of chloroform drops, "galvanisation" (application of a weak electric current), etc.: in short, his method consisted of "*Variatio delekat*"—i.e., medical entertainment.

Now that he had Freud's mother firmly in his grip, Stekel made the most of the "case" and reported, in a flow of letters to Freud, about the condition of "dear mother," at whose steady disposal he declared himself. She was constantly "improving," yet had "bad attacks." On 14 July he reported: "Your sister reveals nice conversion symptoms. She was going to tell Steinsberg that he need not return, as Reitler is taking over treatment. But she did not have the courage to do it. Instead, she got a terrible sore throat, and Dr. Steinsberg had to come to examine her throat and found ... nothing. The words got stuck in her throat and she wanted Dr. Steinsberg to get them out (to guess it) ... She cleared her throat and asked: You really see nothing? In the evening she came to me: I can't find the words ... How shall I say it to the doctor? Why doesn't he understand himself?"

A typical example of Stekelian "intuitive" analysis that was no doubt meant to impress Freud. It was certainly either courage or impudence that drove Stekel to take Freud's mother hostage and to psychoanalyze the family; the effect, however, was not a re-strengthening of his relationship with Freud, as he no doubt had hoped, but the reverse.[1] On 15 September he travelled back to Vienna to find an unpleasant surprise waiting for him: Viktor Tausk, a young and ambitious analyst presented himself as a proposed member of a newly founded "review committee" of the *Zentralblatt*. Stekel considered Tausk to be his "arch enemy," someone whose presence he could "hardly bear in the society." In May earlier that year he had demanded from Freud that "Dr. Tausk gives you the solemn promise that he will not mention my name in this society at all any more and refrain from entering into discussion with me," if not he would be forced "to draw the ultimate consequences."

With Tausk dropped on his doorstep, Stekel was now clearly invited to draw the ultimate conclusion. Indeed, objecting strongly to Tausk as his new co-worker, Stekel wrote to Freud on 14 October 1912: "I find it unpleasant to tire you with these internal troubles, but I have been pushed into this position, and reconciliation [with Tausk] now or later is out of the question. We need all the forces we have, now more than ever ... I do not want to pick a fight, only some consideration of the present situation. *Frankly: every review from Tausk in my journal would spoil for me the pleasure of working on my beloved Zentralblatt*" (emphasis added).

This, we believe, must be considered the actual "refusal" of Stekel to comply with Freud. Freud, at any rate, chose to interpret this sentence as an insistence on Stekel's part on managing "his" journal without interference from anyone. He presented it as the immediate reason to break off all relations with him, even though in fact Stekel claimed no more than that his pleasure in the journal would be spoiled if Tausk were to publish in it. In fact

[1] Stekel did not necessarily initiate the psychoanalysis of the Freud family: Alexander Freud's wife Sophie was sent to Adler for an analysis.

Stekel later claimed to Jung that at this point he offered Freud to resign from the *Zentralblatt* if "things could be settled in a friendly matter."

Significantly, the next item in the Stekel–Freud correspondence is a note from Stekel to a certain Herr Rat, possibly a solicitor who was consulted by Freud or someone who was charged by the Freud family to handle a financial transaction, namely the payment of Stekel for his services rendered to the mother.[2] Rat must have been instructed to act on or very shortly after the day Freud received Stekel's letter of 14 October, since the note to Rat is from 3 days later. Stekel refused payment: "The allusion [to payment] alone is an affront. Surely I can do the mother of my teacher a small favor?" he wrote back. Someone underscored in pencil the words "allusion" and "affront."

Another note from Stekel to Rat, again dated only 3 days later, makes it clear that the other party does not take no for an answer: "I have never thought about [asking] a fee and never booked the number of visits I made to Frau Mama [Freud]. Next time when my assistance is called upon I shall assuredly do so, and I herewith consider the business concluded". And in an aside Stekel wrote to Rat: "In confidence! I can probably settle matters with the brother [presumably Alexander Freud]. He seems unable to recognize and *keep* his true friends ... ". *He*. That would have to be Freud.

Written across the entire letter someone (but not Freud) angrily wrote in pencil the word "swine" twice. *SCHWEIN*! The same hand also placed two exclamation marks before the "confidential" passages and underlined the sentence "in confidence." At the bottom of the page, finally, a handwritten calculation mentions the payment of 300 Kronen.

In a last note to Rat, dated 23 October 1912, Stekel gives in. "I cannot accept such a large fee for those few consultations. I will keep, so as to reassure your conscience, 50 kronen and return the remaining 250 with thanks."

Indeed, the conclusion of their business now approached its end. From Andreas-Salomé's diary (1987, p. 35) we learn that Stekel, who realized that he was cornered, sought contact with Adler once more, but Adler refused to be "drawn into the fight that flared up." On 27 October 1912 Freud wrote to Ferenczi to explain the situation: "We shall get rid of Stekel, which is worth a sacrifice in itself." He mentioned how that summer, after their last reconciliation, Stekel had begun his quarrels and jealousies anew, and how Stekel suddenly declared that he could never accept Tausk on his journal, emphasizing the word "his." In retrospect, Stekel thought that Tausk had been used as a "battering ram" to crack "fortress Stekel." That seems in fact to be an accurate evaluation of the situation, for "Tausk" probably *was* the occasion Freud created to get rid of Stekel. The *Zentralblatt* affair was, as is well

[2] It is possible that by "Rat" not a name was meant but a (perhaps ironic) form of addressing, since "Rat" means "advice" in German. In that case, Freud and Herr Rat may have been identical.

known, given as the "official reason" for the definite break between Freud and Stekel. "His [Stekel's] behavior, of which it is not easy to publish an account, had compelled me to resign the management and hurriedly to establish a new organ for psychoanalysis" (Freud, 1914, p. 47): words that have been quoted and interpreted many times to probe secret meanings behind them (for example Kuhn, 1998; Rudnytsky, 2002). But "Tausk/Zentralblatt" did not cover a deeper meaning. It was not, it is true, "anything ultimate," nor was it a "real agent" (to paraphrase the masturbation dispute between Stekel and Freud): it was the conclusion of an argument that had already been started perhaps ten years earlier.

Dialects of Estrangement: the Dynamics of the Freud–Stekel Rupture

A long time may have elapsed between the moment that two people start to work together to the moment that it has become impossible for them to continue doing what they did before, and many incidents may have occurred, many events may have influenced their lives. Yet when we want to understand why they decide to discontinue their working relationship, we are by and large looking for a *particular* event, or *series* of events, that "caused" this break. And this is the case here too, where an early schism is understood as a gradual but steady deterioration of a relationship of two incompatible characters. The final result is a necessary and unavoidable total collapse.

Plotted against the axis of time, we might imagine a tortuous declining line, slowly approaching zero. But this kind of representation of things is illusory in two respects. First it does not include the fact that this is a reconstruction with hindsight. The direction of the arrow should point backwards, not forwards. Thus on 31 July 1912, when Stekel is less then three months away from his final break with Freud, he writes: "You know how much you mean to me and the part I play in your history. In the history of psychoanalysis, our names are closely linked to each other, and I remain the prophet, you the Allah!" (a phrase he would use again in his 1926 historic narrative and his autobiography, when he declared himself to be the prophet of Freud, his "Christ"). Was Stekel completely ignorant, or did their break occur so suddenly that he was taken by surprise? Neither, we believe. The desperate doctoring of "dear mother" suggests that Stekel was well aware of the fragility of his relationship with Freud, and that he tried to negotiate a better position for himself.

Similarly, in his letters to others, Freud often scorned Stekel and openly declared that he was looking for an occasion to get rid of him, but each time that Stekel had fallen out of his grace a reconciliation had taken place which restored Stekel to his position of rebellious but original follower (as he no doubt was wont to see himself). There were no compelling reasons why the same might not happen this time too. And indeed, Stekel's post 1912 letters give the impression that he never lost hope that a reconciliation could not take place again. (Fritz Wittels, a follower of Stekel, did reconcile with Freud after having been out of touch for many years. Wittels tried to reconcile Stekel with Freud—to no avail, see Timms, 1995, p. 121–122.)

All this suggests, we believe, that there was not a progressive deterioration in their relationship, but on the contrary: a relationship that was founded on such controversies. The problem with the theory of deterioration is that it is "undialogical." It treats discursive processes such as negotiation, argumentation, dialogue etc. in the same manner as it treats history: roughly as the by-products of a specific interpersonal dynamic, rather than as instruments *in* this. But the pragmatic effect of speech is not so much in what other people do (or do not do) in response to one's utterances, but in how one estimates what other people will do. Language usage presupposes "mind reading" of a kind says Carrithers (1992), and the successful speaker is the one who is able to fine tune his thoughts to those of the other. But note that fine tuning is not identical with reaching agreement: fine tuning is adjusting discourses to one another.

In our reading of this correspondence, we emphasize the negotiation of contested meaning in dialogue. Recall how we argued in chapter 5 that what constituted the (sometimes covert) dialogue between Freud and Stekel was the latter imitating the former at various levels. Through imitation (after all, the oldest form of learning) Stekel found a way of challenging Freud, which Freud did find, at least to a certain degree, appealing. There is a resonance of this sentiment in an aside in Freud's letter to Stekel of 31 January 1924 when he wrote: "Perhaps you remember how our relationship began." Notice: no question mark. A dialogue between the lines.

But in this, in the imitation, Stekel also sowed the seeds of their later fracture. In the correspondence that we have at our disposal, we can see, even without having access to the larger part of Freud's side, how the process of fine tuning begins to give way; or better yet: is turned upside down. Style, tone of voice and choice of subject matter may have differed in fact little from other minor correspondence (Groddeck, Pfister, Weiss and others), but it clearly worked against Stekel more than it worked against others.

Thus the wavering between various forms of address (*Sehr geehrter Herr Professor! Lieber Meister! Lieber Herr Professor!*—never without an exclamation mark) already seems to indicate a certain lack of coherence; certainly his all too frank observations and comments about others suggest a lack of distance. With the "analysis" of the Freud family he definitely crossed a barrier ...

Similar are the "priority problems," as Stekel preferred to call them: questions that related to ideas he claimed as his property (the discovery of a new symbol; the priority of having first written on "déja vu," etc.), which always raise the issue of mutual dependence: I learned from you, *but you also learned from me*! So, more particularly, the demand that his voice be acknowledged and his contributions be referred to if only in private—but preferably through disguised allusions. Thus in an undated letter from 1912 he draws Freud's attention to "a little violet I claim to be mine. Not in public! God forbid," and elsewhere the demand to be acknowledged as a "necessary counter voice" in psychoanalytic discourse. In all these claims there is the suggestion that he and Freud are on the same level, that they contribute to each other's development, yes, even that they are equally interested in each other. "I only received the *Jahrbuch* just now, and I am instantly and greatly impressed by your paranoia work (Freud, 1911). Many of the new ideas were already present in me as seeds, as you could make out from my little work on exhibitionism. Nevertheless, it is so new and overwhelming that I must speak to you about it."

It is interesting to note that Isidor Sadger made the very accurate observation in his recently recovered *Recollecting Freud* (2005, pp. 39–40) that Stekel "tried to put himself up as an equal to the Professor on the basis of his own little discoveries"—and that Freud did not tolerate deviances from his teachings, the less so with students more close to him. "Collegiality expands with the square of distance," Sadger wrote (2005, p. 38), not realizing that Stekel used a very similar expression in a letter to Freud.[3] Sadger and Stekel strongly disliked each other, but their historic accounts are eerily similar in many respects.

The suggestion of sameness, of equal intellectual level, can be traced back to Stekel's close imitation of Freud's style which played a constitutive part in his becoming a psychoanalyst. By 1910 however, those very same discursive constructions turned against him. Instead of being read as "provocative" in the positive sense of the word, all these improper claims to priority were now interpreted as delusions of grandeur. Neither reading was ever absent from Stekel's discourse, but the one was always suppressed in favor of the other. It is in compliance with the discursive dynamic described here, that that these frames of interpretations now switched. With their relations cut off, each letter that he wrote to Freud would be filed in his dossier as one more proof of his estrangement. This new frame of reference explains why Stekel's position would be rewritten in psychoanalytic history retroactively as always having been marginal anyway. It is in this that Stekel was and will forever be "marginalized."

[3] Stekel to Freud, 1.6.1912: "Nothing alienates more than proximity ... "

Chapter 8

ON THE HISTORY OF THE ANALYTICAL MOVEMENT[1]

Dr. Wilhelm Stekel Translated and annotated by Jaap Bos

Our memory is partial. Freud proved this and presented numerous examples from which this conclusion follows. When I present an account of the history of the analytical movement, I can assert that such a "partial functioning of the memory" has not strongly developed in me. I have no loss of memory. I see my first childhood experiences in all their details clearly before my eyes. Early on I made it a habit to face the truth fearlessly, when I fail, I blame myself. I am not vengeful and I find it easy to forgive because I have learned to appreciate what is human in many responses, and this was the case even when psychoanalysis was still unknown to me.

When I try to write such a history of the analytical movement, I have no other motive than to serve the truth and to answer attacks—open as well as covert—the intention of which is to undermine my researches.

I was a sex researcher long before I became acquainted with Freud. Already as a Gymnasium student, I had examined the first sexual experiences of my friends, and later, when I was a general practitioner, I designed a survey in which one was supposed to report one's first sexual experiences. In 1895—independently of Freud, and still unaware of his teachings—I published my study on "Coitus in Childhood" [Stekel, 1895]. About a year later in the Vienna general hospital I met the brilliant internist Dr. Max Kahane, who died a few years ago.[2] We were both looking at a notice that

[1] This chapter first appeared in *Psychoanalysis and History*, 7(2005), 99–130. All footnotes are the translator's.

[2] Max Kahane (1866–1923), Viennese internist, befriended Freud from an early age. He attended Freud's lectures at Vienna University during the semester 1895–1896. Kahane was a member of the Vienna Psychoanalytic Society from the very beginning, in 1902, until 1907. He was director of the Institute for psychical treatment in Vienna from 1901 onwards. The reasons for his break with Freud are unknown. He took his own life in 1922. Among his main publications is *Medizinisches Handlexikon für praktische Aerzte*. Berlin/Vienna, 1908.

announced upcoming lectures. Kahane drew my attention to a course by the lecturer Dr. Sigmund Freud. "This should interest you as a sex researcher." I resolved to attend the lectures. However, I did not manage to get to them. I was a very busy practitioner, with many interests and I was continuously attempting to expand my medical knowledge, and my neurological knowledge in particular.

At this time, a weekly periodical was published in Vienna under the auspices of Hermann Bahr, Singer, and Kanner. Reading *Die Zeit* [The Times] was a sign that one was highly educated. And this is where I came across an article by Max Burckhardt which ridiculed a book written by Freud: *The Interpretation of Dreams* [Freud, 1900].

Up until then I had never given the problem of dreams a thought. It had only occurred to me that the dream revealed forbidden sexual meanings of which I was not aware while awake. In these wet dreams, the sexual nature of dreams came to the fore (as a forbidden wish), and this was something that had entirely escaped my consciousness. I firmly resolved to pay Freud a visit.

The opportunity would present itself soon. Confusions made it seem well-advised to contact an experienced sex doctor. I paid my visit to Freud, who had quoted my brief study on "Coitus in Childhood" in his work *The Aetiology of Hysteria* [Freud, 1896].[a] I was therefore not a stranger, I could freely put my discomfort to a sympathetic ear, and seek advice and help. Freud proposed psychoanalysis, which took approximately eight sessions. I told him my very interesting childhood experiences. Freud wondered about the fact that I had no repressions. We also conducted dream analysis, and Freud aimed to demonstrate a fixation I had with my mother. I was stunned. My leaning towards my mother was that of a normal person; I had been able to free myself from her at an early age. I was aware of two incest dreams in my youth. I was not aware of a connection between my sexual life and my mother, who lived miles away. And this is a position I still maintain today. The so-called Oedipus complex is a normal incident. Only an excess makes it a pathological phenomenon. I pointed out to Freud that I had a different opinion, and he replied jokingly:

> "That's life! With this drink in my body, I see Helen in every wife."
>
> I retorted with an old joke:
>
> "Look who's talking, I still see in every wife my mother."

We both laughed, and the hour passed in merriment. Freud even emphasized what an exceptional human being I was: I had no repressions, and this alone made me a valuable witness to infantile sexuality.

I now began to study *The Interpretation of Dreams* and became quite enthusiastic about it. In 1901 I wrote a feuilleton for the *Neue Wiener Tagblatt* in two lengthy parts, entitled "Traumleben und Traumdeutung" [Stekel,

1902], and I was the first person who dared to defend Freud and his great discoveries.[3]

I even stood up bravely for the master in German newspapers (*Vossische Zeitung* and *Frankfurter Zeitung*). Freud gave me a copy of his *The Interpretation of Dreams* with the inscription: " Herr Dr. W. Stekel with heartfelt appreciation for reviewing this book with integrity." On 23 October 1902 I was in a position to review *The Psychopathology of Everyday Life* (Freud, 1901), which at the time was published in the *Monatschrift für Psychologie und Neurologie*. I was tireless in my efforts to advocate Freud. Wittels was not wrong when he said: "The printing machines in Europe sighed deeply under the weight of work in which Stekel defended Freud."

I cannot recall exactly when I began to analyse myself. It may possibly have been in 1903. I knew that a certain Dr. Reitler had already tried his hand at psychoanalysis.[4] I believe he was the first to attempt to apply Freudian teachings in practice. He was an enthusiastic follower of Freud, and he remained so until his death.

My first big case—I had already conducted a number of smaller analyses in my private practice—was referred to me by Freud. It was the case of a rabbi who I described extensively in the first volume of my "Disorders [of the Instincts and the Emotions]" series (Chapter 12, case Nr. 112, page 336 in the 4th edition).

When I think of it in retrospect, I believe it was in fact a risky enterprise. I could have seriously messed up the case. Every two weeks I went with my patient to Freud, but otherwise he was entirely left in my care with no further supervision. Freud trusted me implicitly—or so it appeared. However, I found out later that he also did the same with less talented pupils. He would refer a case to them without providing the guarantee of a control analysis. By this time I had already read Freud's *The Interpretation of Dreams* [1900] and the *Psychopathology of Everyday Life* [1901]; I read the *Studies on Hysteria* (Freud, 1895b) much later. Freud said it was outdated and anyway I knew everything there was to be learned from it.[b]

My first case went surprisingly well once a certain amount of resistance had been overcome. It ended in complete success. From now on I was more of an analyst than a general practitioner. At last I had found the method to

[3] Jones (Vol. II, p. 7) wrote that Stekel incorrectly suggested (in his *Autobiography*) that he published his review of Freud's *The Interpretation of Dreams* in response to Burckhardt's review, published in *Die Zeit*, and that in actuality he wrote his much later, "presumably after his analysis." From Stekel's 1926 account we now learn that this was in fact true.

[4] Rudolf Reitler (1865–1917), Viennese medical doctor, also befriended Freud from an early age. Member of the Vienna Psychoanalytic Society from 1902 to 1914. Reitler worked in a health resort in Baden, near Vienna, and also had a private practice. Was known to be the first to practice psychoanalysis after Freud. Fell ill in 1914 and died three years later. Reitler wrote a lengthy book-review of Stekel's *Nervöse Angstzustände*, praising Freud for having pointed out ways to approach psychotherapy (Mühlleitner, 1992, pp. 266–267).

open up the mental life of my patients. Earlier on, I had already begun to speak to patients independently of Freud, and to interest myself in their lives and finally tried my luck with Dubois.[5] But only now did the world of neurosis reveal itself to me. I visited Freud more often, reported my findings to him, and advocated him and his new teachings both in writing and in speaking. Finally, I suggested he start a small circle comprising medical doctors, literati, and artists who would discuss psychological matters and also his new teachings. He was initially enthusiastic, but then worried about obscurity and ignorance, and suggested that I start a small esoteric circle for the esoteric to join in. He had a few friends and followers, and he wanted to call them together in regular meetings.

So the first gathering took place in the evening at Freud's home. Apart from the master and my humble self, the following medical doctors were present: Dr. Rudolf Reitler, Dr. Alfred Adler, and Dr. Max Kahane. Freud initiated a specific ritual which continued for many years. We arrived there after the evening meal. There were cigars on the table to which most of the participants helped themselves generously. In the beginning Freud smoked almost uninterruptedly. He smoked a small English pipe for which he had sacrificed his beloved Trabuko cigars, only to return to Trabukos again at a later stage. I have rarely seen a man smoke so much. (He had also given me an English pipe and English tobacco. Smoking must have been a complex of the master. From what he told me, I had the impression that he had "oversmoked" himself. So he had to quit smoking for a few days.) Later black coffee was served by one of the gentlemen.

It was only natural that we hit upon a psychological theme after a few introductory remarks. It was Kahane who proposed that we speak freely on different subjects, and he initiated a conversation on smoking.

The next morning I wrote the conversation down and published it in the *Prager Tagblatt*. This conversation is presented verbatim below simply because it gives a good idea of our first unconstrained conversations.

Conversation on Smoking

A small, cosy study belonging to an important neurologist. The owner sits at his writing table and smokes his little English pipe. The "tireless spirit" leans

[5] Paul Dubois (1848–1918), Swiss Psychiatrist from Bern, developed a method of psychotherapy called "persuasion." Through reasoning, suggestion and persuasion, Dubois challenged his patients to develop a rational perspective on their situation, which would aid them in acting decisively to address the issues that made them unhappy. Important publications include: *Les Psychonévroses et leur traitement moral: leçons faites à l'université de Berne* (1904). Paris: Masson et Cie, Éditeurs, 1904 (Translated as *The Psychic Treatment of Nervous Disorders: The Psychoneuroses and Their Moral Treatment*. Translated and edited by Smith Ely Jelliffe and William A. White. New York/London: Funk & Wagnalls Company, 1907).

back in a comfortable armchair and, like his master. smokes an English pipe in a similarly relaxing manner. The "Reticent" deals with a fine cigarette with considerable expertise and elegance. The "Socialist," relaxed, draws on a Virginia while adopting a somewhat serious demeanour.

The doorbell rings.

In comes the "casual." The landlord offers him a cigar.

THE CASUAL: "No, thank you. I smoke very little at present. I have come to the conclusion that I can think much more easily if I don't smoke. Smoking produces a certain sense of well-being. It stimulates one's creative spirits. But it has one big disadvantage. Just like alcohol, it robs us of a clear view, it destroys the beneficial influence of self-criticism."

THE MASTER: "I see. You wish not to deprive yourself of one atom of your free will by smoking."

THE CASUAL: "That's right. I now work much more clearly and lucidly. But I would like to hear from the other gentlemen what they think of this. Could it not be that the decline of metaphysical sciences, the withdrawal of philosophy from the other sciences, be attributed to the general practice of smoking when doing intellectual work?"

THE TIRELESS: "That's not my experience. When I think back, I find that my writing activities began exactly at the same time as I began smoking."

THE CASUAL: "That only goes to prove my premise that smoking undermines the self-critical faculty."

THE LANDLORD: "Witty—but vicious."

THE TIRELESS: "—and wrong. Because I only smoke after dinner, and never when involved in intellectual work. When did the Europeans start smoking? It became general practice only about two hundred years ago. Until the nineteenth century, smoking was not allowed in public. But it was precisely during that last century when the most important discoveries were made, and during which such incredible intellectual works were produced."

THE MASTER (to the casual): "You explore the nature of creativity directly. I, at least, proceed in two stages. During the first day, while still under the influence of my own enthusiasm, I conceive my thoughts. At this stage, fantasy is required if I want to arrive at a general outline. Critique comes to me during the second day. Whether I'm smoking or not. However, your exposition does seem to have an element of truth to it. When I set out to read a rival's book, which annoys me in advance, I never smoke. I read it through quickly with considerable agitation and ... only smoke afterwards. Smoking definitely produces a slight narcosis, a relaxation of the nerves."

THE RETICENT: "There are some interesting observations to make about women who smoke. Some of them smoke to satisfy their emancipatory pleasures ..."

THE TIRELESS: "They smoke because they want to have pleasure the way men do. Why look for deeper sins?"

THE MASTER: "I know of a intelligent girl who is a keen smoker. Asked about it, she defended herself with a passionate poem. Briefly, the gist of it is: 'I smoke so much because I am kissed so little'..."

THE TIRELESS: "That could have a double meaning. The delight in nicotine appears to diminish our want of love. There are cases ..."

THE MASTER: "I know what you are driving at. We all do. It explains the eternal hostility our women feel towards smoking!"

THE CASUAL: "That's hilarious! And the reproach that the smoke gets into the curtains ..."

THE MASTER: "... is but a pretext."

THE TIRELESS: "A smokescreen, that obscures the real link." (all laugh)

THE RETICENT: "It's all perfectly clear now".

THE TIRELESS: "No, nothing is clear. There is something wrong again. There is a girl I once loved..."

THE CASUAL: "What was her number?"

THE TIRELESS: "...Not important...whom I once loved. When I kissed her, she said: You don't smoke? That's really strange. You don't smell like a man. A man ought to smoke. She pronounced that word 'man' in such a solemn way: a man—formally underlined, three times."

THE CASUAL: "How old were you then?"

THE MASTER: "It was just a hidden wish to see you as a grown-up man."

THE SOCIALIST: "After the wedding she would want you to do just the opposite!"

THE RETICENT: "I believe that too."

THE TIRELESS: "That's chitchat, not psychology. I seriously believe that women would not want to miss out on the smell of tobacco in our kisses."

THE CASUAL: "Like with a moustache, which should interfere with the kissing."

THE SOCIALIST: "So smoking has, in many cases, intimate sexual connotations. This can be evidenced by the fact that old spinsters in the mountains are often much worse smokers than men. They have had to renounce love."

THE TIRELESS: "The master is right. Smoking produces a slight narcosis. But we medical doctors have to admit that it can become a dangerous poison as well."

THE MASTER: "In large quantities smoking becomes dangerous to men. But what harm could moderate smoking do? I always become distrustful when I hear that a patient died from smoking too much. You very often find other causes hidden behind it. Most often a combination with alcoholism, which seems to be particularly noxious."

THE TIRELESS: "I actually smoke very little. Five cigars a day, and always after a meal. And I also believe that I could quit at any moment. Indeed—after a particularly bad cigar I say to myself: tomorrow, you'll quit. I have deceived myself for many years with the idea that this cigar was the last one. The worn-out consolation of all sinners! Big or small."

THE MASTER: "You're not a smoker. You don't understand. I was not allowed to smoke for two years. It was horrible. It felt like a good friend had died and I had to mourn him from morning to evening. Now I even have the same feelings for my pipe. She is a good friend of mine, my counsellor, my comfort, my guide, who smoothes my way."

The hour struck twelve. We brought our conversation to a close.

In this short conversation, we have before our eyes the kernel of the analytical evenings. The master is Freud, the socialist is Adler, the casual is Kahane, the reticent is Reitler, and the tireless is myself.

It soon became clear that a specific programme would be required for every evening. One gentleman would present a lecture to be followed by a discussion. Specific rules were gradually formulated, and they were adhered to to the letter. New members joined. First there was Dr. Max Graf, the composer, then Dr. Max Bach, the music critic. (One session—I believe it was the fourth or fifth—has been recorded as a photostatic faithful representation, and was published in my collection of essays *Masken der Sexualität* [Stekel, 1920a], the title of it is "Der Kleine Kohn. Zur Psychologie des Refrains" [Little Kohn. On the Psychology of Refrains]. The starting point was a popular refrain of the time entitled "Haven't you heard of little Kohn?".) Among the medical doctors, Dr. Paul Federn and Dr. Hitschmann joined us, later a certain Mr. Heitler.[6] Next Dr. Sadger, who later introduced Dr. Wittels. One day Dr. Urbantschitsch lectured as a guest, recounting his very interesting life.[7]

The evenings were still held at Freud's and were fairly unconstrained events. Freud presided. The speaker lectured on a prescribed topic, and subsequently names were drawn from an urn in order to establish the order in which people would speak.

The first few evenings were wonderful and left an indelible impression in the minds of every participant. Jealousies and rivalries had yet to surface. Freud's reputation had not yet been made and we had a common goal: to bestow on the master the recognition he deserved.

[6] The identity of Mr. Heitler has not been ascertained; he is not mentioned in either the *Minutes* or in Mühlleitner's *Lexikon*. It is possible, however, that Stekel meant Adolf Häutler about whom little is known. He was an "erudite man, interested in the application of psychoanalysis to philosophic, especially aesthetic problems" (*Minutes*, Vol. I, p. xxxiv).

[7] Rudolf Urbantschitsch (1879–1964), studied medicine in Vienna and worked from 1904 on in a Sanatorium in Edlach (Austria). In 1908 he opened his own sanatorium. Urbantschitsch was a member of the Vienna Psychoanalytic Society in 1908 for a brief period only, during which he gave two presentations. He sold his sanatorium in 1922 to devote his life to psychoanalysis. He was analyzed by Federn and later by Ferenczi; he began to practice psychoanalysis in 1924 but was not a member of the Vienna Society. In 1936 he moved to the United States and worked in Los Angeles and San Francisco. His autobiography "Myself not least" was published in 1958.

At that time I was the only analyst apart from Freud. Reitler was a doctor in a health resort in Baden near Vienna, and was called to work in a resort in Vienna. He did conduct analyses, but not with the same passion that I had done.

Unfortunately, I was to make my first discovery that would cast a shadow on my relationship with Freud. One evening I lectured on "The Psychology of Anxiety Neurosis." Freud believed anxiety neurosis to be an organically determined condition that was brought about by some detrimental form of sexuality. (He later went on to call this condition actual neurosis.) Three carefully analysed cases of anxiety neurosis enabled me to prove that it was caused by a psychological event.

The first case of agoraphobia that came to my practice was that of a cashier, of whom I was able to point out that he toyed with the idea of perpetrating fraud, and of using the money to travel to America (Volume I, 4th edition, case Nr. 1, page 13.). Freud had taught us that psychology is ruled out in cases of phobia. Now here was a severe case of phobia and behind it there was a psychological conflict!

After I expounded a case of anxiety neurosis analytically, Freud said: "That is not a case of anxiety neurosis, that's hysteria!"[c] By way of response I drew a calling card from my pocket. I had sent the same patient to Freud, who had then sent her back to me with the diagnosis 'anxiety neurosis.' Freud looked at the card in silence and then said: "Do not publish this case before we have discussed it."

I received an invitation to visit him on Sunday. Freud never analysed on Sundays, he only read, wrote, or received his friends.

So I visited Freud on Sunday. I can still clearly recall our conversation. Freud: "I have given your analysis some considerable thought. I want to give you a royal present. I propose classifying the cases you have described as 'anxiety hysteria,' and to make a sharp distinction between that and anxiety neurosis. Anxiety patients with a psychological root are anxiety hysterics, whereas those whose condition is determined by a detrimental form of sexuality are anxiety neurotics."

I stuck to this classification at the time, even though I could never agree to it entirely. In my 1907 study, published in the *Medizinische Klinik* under the title "Nervous anxiety," as well as in my simultaneously published pamphlet *Ursachen der Nervosität* [Causes of Nervousness, 1907a], I laid down the imperative to which I have remained faithful to this day: "*Every neurosis is determined by a psychological conflict*."

In the introduction to my pamphlet, which was dedicated to Freud, I wrote: "I attempt to prove that deep down, all conditions which both doctors and laymen call 'nervousness' point to a mental illness. I have come to this conclusion through lengthy, intense involvement with the psyche of the nervous. But it was Prof. Sigmund Freud, the great soul expert, who has shown me the way, and to whom I owe the new light that shines in the darkness. I proudly concede to being his pupil, which is not to say, however, that *everything*[d] which I maintain here is in line with his views. On the contrary! I'm quite sure he will shake his head at many of the sections, as if denying his approval."

I was to prove right. Freud shook his head energetically, and to this day has refused to acknowledge that all neuroses are psychically determined. In his recently published, fascinating autobiography [Freud, 1925], which is even more interesting with respect to what it does not say than to what it does say, he still clings to the theory of actual neurosis (anxiety neurosis and neurasthenia). As before, he makes a distinction between actual neuroses and anxiety neuroses. (Freud's psychoneuroses included at the time only hysteria and compulsive neuroses. Phobias were still considered to be organically determined. He is in debt to me for this advance in knowledge.)

Freud has a character trait that has done him great harm and has slowed down the development of analysis, and yes, even taken it in the wrong direction: *He is unable to accept that he erred*! He isolates himself from all corrections of his established viewpoints. (And this man is supposed to understand his opponents, who equally isolate themselves, and prefer to steadfastly maintain what they at one point believed to be true.)

But what is scientific truth? What is left of what was, only a few decades ago, state of the art? And what will remain of our knowledge today in a hundred years' time? Thus Freud persists in his errors, speaks of actual neurosis and psychoneurosis, from a differentiation between neurasthenia and anxiety neurosis, etc. In his autobiography [Freud, 1925, pp. 24–26] he literally says:

> Under the influence of my surprising discovery, I now took a momentous step. I went beyond the domain of hysteria and began to investigate the sexual life of the so-called neurasthenics who used to visit me in numbers during my consultation hours. This experiment cost me, it is true, my popularity as a doctor, but it brought me convictions which to-day, almost thirty years later, have lost none of their force. There was a great deal of equivocation and mystery-making to be overcome, but, once that had been done, it turned out that in all of these patients grave abuses of the sexual function were present. Considering how extremely widespread are these abuses on the one hand and neurasthenia on the other, a frequent coincidence between the two would not have proved much; but there was more in it than that one bald fact. Closer observation suggested to me that it was possible to pick out from the confused jumble of clinical pictures covered by the name of neurasthenia two fundamentally different types, which might appear in any degree of mixture but which were nevertheless to be observed in their pure form. In the one type the central phenomenon was the anxiety attack with its equivalents, rudimentary forms, and chronic substitutive symptoms; I consequently gave it the name of *anxiety neurosis*, and limited the term *neurasthenia* to the other type. [reference omitted] Now it was easy to establish the fact that

> each of these types had a different abnormality of sexual life as its corresponding aetiological factor: in the former, *coitus interruptus*, unconsummated excitation and sexual abstinence, and in the latter, excessive masturbation and too numerous nocturnal emissions. In a few specially instructive cases, which had shown a surprising alteration in the clinical picture from one type to the other, it could be proved that there had been a corresponding change in the underlying sexual regime. If it was possible to put an end to the abuse and allow its place to be taken by normal sexual activity, a striking improvement in the condition was the reward.
>
> I was thus led into regarding the neuroses as being without exception disturbances of the sexual function, the so-called "actual neuroses" being the direct toxic expression of such disturbances and the *psychoneuroses* their mental expression. My medical conscience felt pleased at having arrived at this conclusion. I hoped that I had filled up a gap in medical science, which, in dealing with a function of such great biological importance, had failed to take into account any injuries beyond those caused by infection or by gross anatomical lesions. The medical aspect of the matter was, moreover, supported by the fact that sexuality was not something purely mental. It has a somatic side as well, and it was possible to assign special chemical processes to it and to attribute sexual excitation to the presence of some particular, though at present unknown, substances. There must also have been some good reason why the true spontaneous neuroses resembled no group of diseases more closely than the phenomenon of intoxication and abstinence, which are produced by the administration or privation of certain toxic substances, or than exophthalmic goitre, which is known to depend upon the product of the thyroid gland.
>
> Since that time I have had no opportunity of returning to the investigation of the "actual neuroses"; nor has this part of my work been continued by anyone else.

In these quotes we encounter the entire Freud. He did not budge an inch from his initial viewpoints. Being an actual neurosis, anxiety neuroses continue to be brought about by a disruption in the sexual life; neurasthenia still exists as a special type of illness, even though Freud no longer encounters it. His material has changed. It is not the way he perceives illnesses that has changed, no, it is simply that the material has changed. He has not seen a case of neurasthenia for 20 years. Freud does not want to admit that everything which was described as neurasthenia up until now he now calls anxiety neurosis, apart from cyclothemia and early schizophrenia.

Again, the detrimental effects of excessive masturbation are mentioned. Freud could not forgive me for pointing out the harmlessness of masturbation, and for having proven wrong his conclusion that neurasthenia is brought about by masturbation. Stekel upset all these truths, but now they are called true again and that is sufficient.

While I was still a belligerent member of the society, I once offered a significant prize. I said I would give a prize to any one of the gentlemen who could show me a case of neurasthenia in which I could not find a mental conflict. Freud says his material has changed. And this explains the following sentence in his autobiography: "Since that time I have had no opportunity to return to the investigation of the 'actual neuroses." True! But what about the claim: *"Nor has this part of my work been continued by anyone else?"* I consider that to be serious ingratitude on the part of the master toward his pupils. I have continued these studies. In a major work I pointed out the importance of the influence of the psyche.

Freud retreats to the organic, to the physical constitution, with which he protects all weaknesses in his teachings. Can he be certain that withdrawal symptoms following a morphine addiction do not signify the outbreak of a neurosis, covered by the narcotic? Does he not know that many of his students are of the opinion that Basedow's disease is psychically determined? And if he is already unable to remember Stekel, he should not forget his English favourite Jones, who has turned out to be very sceptical when it comes to the actual neuroses.

But Stekel is not a scientific researcher, *anathema sit*! [may he be cursed]. Only Freud knows that the development of psychoanalysis can, in many practical respects, be attributed to myself. Has he really forgotten the book I wrote (*Nervöse Angstzustände und ihre Behandlung* [Stekel, 1908])?

I have something to say about the conception and development of this book that is worth recording. After I published my short study on anxiety, I decided to write a larger volume on this condition. Urban the publisher (of Urban and Schwarzenberg) immediately agreed and a contract was drafted and signed within a matter of days. I set to work immediately.

Freud, fearing dissenting viewpoints, offered to write an introduction on condition that I would submit the work to him for correction. I agreed; now the tiresome work began. I worked on one chapter all week, read it to the master on Sunday, and he would cross out whatever he considered to be unsuitable, change other parts, and occasionally make a few minor suggestions. (Freud would gladly share his minor ideas with his pupils, while not imparting his more important ones to anyone.) Naturally everything was now painted in a Freudian manner. Anxiety neurosis made its appearance as an actual neurosis. Coitus interruptus was an angry demon who disturbed happy marriages and destroyed man's robust condition.

I am proud to say that in those days I developed the entire casuistry of anxiety neuroses, and I offered Freud many viewpoints of which he had not been aware until then. I would have to go through it all chapter by chapter if I wanted to show all that I have accomplished. I cannot, and I do not want to decide now who has benefited the most: the master or the pupil. Show me one work where the language of organs is painted in a more creative way with respect to anxiety conditions!

The reading sessions lasted several months, then the more difficult Sundays arrived. I had finished the anxiety neurosis part and was now to start with anxiety hysteria. I could not find the right words—me, who had never had any difficulty writing. Freud had written the first part of the chapter on anxiety hysteria himself and gave it to me. Regis voluntas suprema lex discipuli! [The pleasure of the King is the ultimate law of the disciple.] In the third and fourth edition of "Anxiety States" I returned to the master what rightly belonged to him. His viewpoints have been separated from mine now.

While Freud was very enthusiastic about the first part of the book, I noticed a certain reserve when I began with the larger analyses. Finally he said that these analyses would damage the book; I should confine myself to publishing the first part (the anxiety neuroses) and postpone publishing on anxiety hysteria for a few years.

I was now faced with the prospect of publishing only part of my work. But I was particularly proud of the large analyses and the discovery that anxiety is not converted libido, and I wanted to publish this part at all costs. However, I was financially dependent on Freud, like the whole gang that comes from his club these days. Defending himself against the accusation of intolerance, Freud wrote in his autobiography [Freud, 1925, p. 53]: "But I think I can say in my defence that an intolerant man, dominated by an arrogant belief in his own infallibility, would never have been able to maintain his hold upon so large a number of intellectually prominent people, especially if he had at his command as few practical attractions as I had."

The truth is that a large number of his most important followers are dependent on his benevolence, or otherwise they do not work in a medical setting. Ever since the important doctors (Jung, Adler, Stekel) seceded, or to put it more aptly: were repudiated, philosophers such as Rank and Reik, and lawyers such as Sachs, have become the main pillars of the doctrine, and now practice psychoanalysis.[8] I could give a long list of names of analysts who live off Freud and his reputation! How many patients would entrust themselves to the hands of laymen if they had not been sent by Freud? More about that below.

I was thus financially dependent on Freud, as were all the other members of the circle. He owned our calling cards and bestowed his patients on us as he saw fit. I should like to see the Society the moment this financial cornucopia dries up! Even from abroad Freud was asked to confer reputations on his students, and he was in a position to reward or punish even the remotest of followers.

[8] Otto Rank (1884–1939) studied literature and philosophy, Theodor Reik (1888–1969) studied philosophy and wrote the first psychoanalytic doctoral dissertation (1912) and Hanns Sachs (1881–1947) was a lawyer. All practiced lay-analysis, but only Sachs was a training analyst. All contributed significant studies to nonmedical application of psychoanalysis.

In brief, I was financially dependent on him, I loved him at this time, despite his errors, and I admired his supreme genius as I do to this day. What could I do to save the second part? I took refuge in an old stratagem. I informed my publisher and had him write me a letter in which he demanded publication of the entire book, including the part on anxiety hysteria. I gave this letter to Freud, who shrugged his shoulders and said: "Then there is nothing we can do about it. But you shall see how the critics will carp at you."

The book was complete. All that was still missing was the final chapter. Easter (1908): the first analytical congress (in fact an unconstrained gathering) took place in Salzburg. A few months earlier I had read a little booklet by Jung: *Zur Psychologie und Pathologie sogenannter occulter Phänomene* [1902]. To my astonishment there was a sympathetic reference to Freud in it. I took the booklet to our society and read out the sensational passage. One month later I came across a sympathetic review of it [Jung's booklet] by Bleuler in the *Münchener medizinische Wochenschrift*, who pointed out that Freud had lifted the curtain of our consciousness. Freud was enraptured. Relations between Vienna and Burghölzli were beginning to take shape. Now the followers were to meet. The assembly took place at Easter in 1908 in the breathtaking city of Salzburg. There I met Bleuler, whose work *Affektivität, Suggestibilität, Paranoia* [1906] I had studied with great interest, as well as Jung, Jones, Ferenczi, Brill, Riklin, and the highly gifted Otto Gross. Freud spoke on doubt, and defined doubt as doubt in love. I spoke on anxiety neurosis. Sadger spoke on the psychological roots of homosexuality. It was a truly unforgettably wonderful time. There was not a single sign of the subversion that was soon to set in. Gross' speech was formidable. He was at that time a lecturer in neurology in Graz. Freud was put on a par with Nietzsche and hailed as the trailblazer of a new morality. (At that time I had not read a single sentence by Nietzsche, except for a few aphorisms, and I resolved to study him. Only much later did I understand how much the philosopher and the medical doctor had in common. Nietzsche came to many conclusions which Freud would also discover without having read Nietzsche's works. Freud shared my aversion towards philosophy. In the last few years he appears to have come closer to Schopenhauer.) Brill was the first American who engaged with Freud, but he was not a true-blue American. He is a smart, sympathetic fellow. Ferenczi was the first Hungarian to defend Freud. Abraham, at the time a student of Bleuler, and later the leader of the Berlin Freudians, was also present, and also Riklin, Maeder, Binswanger, and almost all the Viennese followers.

Freud mentioned the fact that a work on anxiety hysteria from my hand was to appear. I had to answer many questions about that.

On his way home Freud read the last chapter of "Anxiety States" and said: "It's very well written." He was in a very good mood. We all looked to the future with optimism. We were after all no longer isolated. It had been decided that the *Jahrbuch für* [*psycho*]*analytische* [*und psychopathologische*] *Forschung* would be published. Freud and Bleuler supported its publication, with Jung as its editor!

Back in Vienna I was to experience my first bitter experience. I had finished my book. The last quires had been printed. Now all that was missing was the preface that Freud had promised me. After much insistence he finally sent me the desired manuscript. I regret I did not keep a copy of it. It would be extremely instructive. But I'm not a collector of historic documents. The preface was written with as much distance as possible, it declined all responsibility for the book, and did not contain any recommendations. I returned it to Freud and said that in the present form I'd rather publish it without a preface. (The publisher, however, insisted on it, but agreed that the first version was simply impossible. It looked more like a critique than a recommendation.)

Freud sent me a second version, which I also had to reject. Only the third version was somewhat more acceptable. I quote a few typical words from it (Freud, 1908d, pp. 250–251):

> Herr Dr. W. Stekel, who was one of the first colleagues to whom I was able to impart a knowledge of psychoanalysis, and who has himself become familiar with its technique through many years of practice in it, has now undertaken the task of working over one topic in the clinical aspect of these neuroses on the basis of my views and of presenting medical readers with the experiences he has obtained through the psychoanalytic method. If I am glad (?) to take responsibility for this work in the sense which I have just indicated, I think it is only right to declare explicitly that my direct influence upon the volume on nervous states of anxiety which lies before us has been a very slight one. (?) The observations and all the detailed opinions and interpretations are the author's own. My share has been limited to proposing the use of the term "anxiety hysteria."

Take note of the cool "Herr Dr. W. Stekel." My copy of *The Interpretation of Dreams,* published in 1900, has the following dedication: "To Herr Dr. Wilhelm Stekel, with thanks for his considerable effort involved in evaluating this book." *The Psychopathology of Everyday Life:* "To Herr Dr. W. Stekel, amicably, cordially"; clearly a degree weaker. Freud referred to me repeatedly as "Herr Dr. Stekel," which was never the case with other authors (such as Adler etc). It is an indication of the distance that was to grow between me and him.

What is also strange is the remark "my direct influence upon the volume [on Nervous States of Anxiety] which lies before us has been a very slight one." We spent at least twenty Sundays working together on it, I had read every single sentence in that book to him, crossed out passages and added new ones. Even the "royal" gift, the expression "anxiety hysteria," had been reclaimed by the master.

Freud had little faith in the success of this book. However, it did turn out to be particularly successful. Even rival journals could not deny the merits of the study. Many reviewers were inordinately positive and wrote of a standard work, a hallmark in psychiatric literature. Jung too was at that time not a Stekel opponent. He wrote in the *Medizinische Klinik*: "Until now, the casuistry of Freudian analysis has been lacking. Stekel's book fills a gap in this respect." Alfred Adler was of the opinion that: "much as Freud's achievements should not be belittled, neither should the works of his followers be denigrated," etc.

This success meant the end of my friendship with Freud and the beginning of a period of anguish within the intimate circle, which had, by that time, grown significantly in size. Freud had predicted a literary scandal. My success made him timorous. He said in our society: "There must be something wrong with this book. Its considerable success makes it suspect." The first edition quickly sold out (1908). It remained sold out for quite some time. Only in 1912 did the second, revised edition appear. And this time I did not have Freud revise it again. It appeared as the first volume of my series on *Disorders of the Instincts and the Emotions*. The preface to the new edition opens as follows: "Readers of the first edition may notice that I have, in many ways, changed my views with regard to the problem of 'anxiety.' The path to knowledge has remained the same, however: psychoanalysis. Neither has anything changed with respect to the therapeutic. Only the way in which I deal with conditions of anxiety may sometimes appear to be new. And yet the germs of these new viewpoints were already present in the first edition. They have ripened in me organically." Why was I so bold as not to submit the proofs to Freud? My book had made me famous and independent of Freud. I had a large private practice, and many analytical cases came to me directly or were referred to me by colleagues.

In this preface I announced a volume on doubt, one on perversions and one on depression. "That should keep me busy for a decade, while it also makes me realize that the difficult and laborious analyses were not in vain. This way I would like to repay the debt that I owe my teacher, Freud."

But I am running ahead of events. In between lay years of considerable struggle and disappointment which would separate me from Freud. The years 1908–1912 mark the development of my autonomy.

My position within the Vienna psychoanalytic society became more and more difficult. Freud no longer had any use for me and was averse to my renewal and improvements of psychoanalysis. If I gave a lecture, almost all the members criticized me *uni sono*, and tore my every word to shreds, only to accept them as fact later on. I was quite upset and did not announce my intention to speak at the next congress in Nuremberg (30–31 March 1910). However, I improvised a presentation on symbolism, which linked up with the previous presentations very well and received much approval, inspiring new research into symbolism. A committee was created—but over the next 2 years it did not receive a single submission. That congress was also to become the starting point of a division among analysts. We had already noticed a long time beforehand how Zurich used to "manipulate" Freud, turning him against the Viennese followers. Freud often told me I had no idea how many times he had to protect me against Jung. I often poured out my troubles to Adler. His laconic answer was: "It's not about you! It's about me—or about both of us." Adler had at that time begun to go his own way. His book on the inferiority of organs, although firmly grounded in Freudian theory, went beyond these limits. Freud remained cool to everything new, and to the two of us. We noticed the influence of Zurich, and it disquieted and worried us somewhat.

Without notifying us, the Viennese, Ferenczi, who had in the meantime become an intimate of Freud's, came to the fore, and proposed an organization for analysts.

His proposals covered the following:

1. Various local societies will be established.
2. Zurich becomes the capital town and takes the lead.
3. An organ of censorship will be created and located in Zurich. All analytical studies must be submitted to this organ and receive Zurich's approval.
4. Jung will be elected permanent President.

The Plan had been brewed up beforehand and was designed to catch the Viennese off guard. Me and Adler were to be silenced.

I expressed my indignation and protested against the proposals. Psychoanalysis should be a free science, a science in development, and one should not try to muzzle it. I pointed out that electing Jung as President "for life" was ridiculous since Freud was the founder of psychoanalysis. If there was to be a "President for life" then it should be Freud. I said: "We in Vienna will never submit to Zurich's yoke!"

Adler spoke with somewhat less emotion but in the same vein. The Viennese were insulted, the German colleagues were on our side, and a straw vote revealed that we were in the majority. In great turmoil, the meeting was cut short for "serious deliberation" and a final decision was postponed until the next day.

I quickly consulted with Adler. At my instigation, we decided to publish the *Zentralblatt für Psychoanalyse*, to appear monthly. We convened a secret meeting of our Viennese colleagues, so as to oppose the proposals and secure a leading role for Vienna.

There were about 15 Viennese present and we were deliberating on counter proposals when the door opened and Freud stepped in. He was ashen and more agitated than I had ever seen him before in my life. (One of the Viennese had played Judas and had informed the master of our secret meeting.)

Freud tried to persuade us and to justify himself. He painted the future of psychoanalysis in dark colours. The entire scientific world would turn against us. Zurich was the only clinic where analysis was taught. An anti-Semitic campaign was being launched against analysis. It was being degraded to a Jewish science. A "Christ" had to lead the movement.

Freud worked himself up into such a state—something I still cannot understand to this day. Tears were in his eyes. "My very existence is at stake," he exclaimed. "They will not leave me the clothes I stand up in."

We were all deeply shocked. We deliberated and finally proposed a compromise, to which Freud was forced to submit. For the first period we would vote for Zurich, but afterwards another society (Vienna) would have to take the lead. Scientific censorship had to be dropped. It had become irrelevant from the mere fact that Adler and I would publish our own journal.

These proposals were adopted the next day by an overwhelming majority. To this day Freud has not forgiven me for being stronger than him at that time. But in one respect he should be grateful to me. I prevented Jung from becoming President "for life." Soon afterwards Jung turned out to be so unreliable that Freud had to take action against him too.

In this historic meeting, Adler and I announced that we would publish a journal which would do justice to the importance of the Vienna society, because the *Jahrbuch für analytische Forschung* was entirely in the hands of the Swiss and through passive resistance prevented publication of Viennese studies. Naturally the opposition was not directed against Freud, but against his Viennese followers (Stekel in particular), who supposedly made analysis a laughing stock in the eyes of the scientific community. That same day the Vienna group was constituted as a formal society. Freud proposed Adler as its chairman, and myself as his deputy. Freud made do with the role of scientific chairman.

Now everything seemed settled again. Freud only worried that we would use the *Zentralblatt* against him. I started negotiations with two publishers: Deuticke in Vienna and J.F. Bergmann in Wiesbaden, who had also published my short study *Dichtung und Neurose* [Poetry and Neurosis, Stekel, 1909b] with whom I had a contract for a another book, *Die Sprache des Traumes* [The Language of Dreams, Stekel, 1911c].

I do not know how Freud came to hear of this. Had I been careless, did he guess right or were there traitors again? I do not know. The fact is, he went to Deuticke and told him that he was prepared to accept the general editorship of the journal assuming that it was run in a way that satisfied him. He wrote a similar letter to Bergmann. Bergmann told me about this and stressed that it would be advantageous for the new journal if Freud were to take on the general editorship. He said he had replied to Freud in the same vein.

The decisive meeting took place in Café Mendl in the Rotenturmstrasse. We met at nine in the evening. Freud opened the negotiations without further ado. "Gentlemen! I pull the strings. Both Deuticke and Bergmann attach great importance to the fact that I am keen to be the editor of the journal. So it depends on me whether or not the *Zentralblatt* appears at all. Now, what guarantees can you give me that the journal will not be used against me?"

Adler retorted: "We don't intend to use it against you at all. The journal is to serve psychoanalysis. It would only be natural if it were used to your advantage."

We finally came to an agreement: Freud was to see all manuscripts. All three of us had the right of veto. Mind you! All three of us! And this veto right needed no clarification. All manuscripts would be submitted to all three of us.

We were in agreement. But Freud had made some smart moves that thwarted our plan to set up an independent journal. The first issue of the *Zentralblatt* appeared in 1911. As a token of our peace agreement with Zurich, the Swiss appeared on the official list of names of co-workers. Freud opened the issue with one study; Adler and I were amply represented. The journal met with general approval and the number of subscriptions increased with every issue.

Interestingly, the first paper to meet with a veto by Freud was a study of mine. It was an article entitled "Die Verpflichtung des Namens" [On the Obligation of Names]. Freud said the article would cause a scandal. I published it in the *Zeitschrift für Psychotherapie* (edited by Moll), vol. 3, issue no. 2, 1911. My paper on doubt met with a similar fate; it was published in the same journal [Stekel, 1912a], but no news of any scandal ever reached my ears.

It became more and more difficult for me to collaborate with Freud. I had discovered the "law of bipolarity" long before Bleuler. This law was mocked and laughed at, and Paul Federn even nicknamed me "Stekel and his bipolarity." To my utter amazement Freud designated ambivalence as the next important step forward in his study on "The Dynamics of Transference." I forced Freud to acknowledge my priority in this respect. In the second volume of the *Zentralblatt für Psychoanalyse*, on page 172, you can read the following characteristic footnote [Freud, 1912a, p. 106 footnote]: "Stekel has proposed the term 'bipolarity' for the same phenomenon."... This footnote, which I insisted be included, does not prevent the Viennese Freudians from speaking time and again of ambivalence, and to attribute it to Bleuler, who discovered it independently of me. I recount these trifles in order to illustrate how petty Freud and his followers were. I had similar experiences with every new idea that I put forward.

My situation became even worse when a member, Dr. Viktor Tausk, was introduced to the society. The maiden speech of this highly gifted but psychopathic person was one of the most painful scenes I have ever witnessed. He spoke on "Kant, Schopenhauer and Freud." In this way he wanted to bridge the gap between philosophy and analysis. In the middle of his lecture the highly excited speaker faltered and was unable to utter another word. Finally I took the floor, to rescue the situation, and I introduced a discussion on the same subject. Tausk burned with ambition. He aspired to be the first Dozent [lecturer] in analysis. He had originally been a judge, but had sacrificed his career to his convictions; he then became a journalist in Berlin and energetically threw himself into analysis. He did not want to analyse as a lay analyst. At the time Freud was still strongly against lay analysis. Rank and Reik were only allowed to work theoretically. Tausk began to study medicine and eventually became a practicing doctor. Unfortunately, his promising career ended in suicide on the eve of his marriage.

I was hated by the man from day one and had to suffer his bitter sneers. Did I stand in the way? He rationalized that he did not agree with my research methods. As long as he opposed me scientifically, I put up with it. Once he got personal and claimed that my case histories were made up.[e] Freud was chairman and remained silent throughout the attack. I wrote him a note saying: "If you don't protect me against Tausk's attacks, I will resign from the society." Freud thereupon mildly reprimanded Tausk and warned him against being personal.

I had just published *Die Sprache des Traumes* [Stekel, 1911c]. *The Interpretation of Dreams* [Freud, 1900] was monumental, but it was just the beginning. Its symbolism was still in its infancy, he had not grasped the importance of death symbolism or religiosity in dreams. With *Die Sprache des Traumes* an important step forward in dream analysis had been taken because it made the analyst independent of the associations of the patients. (Oddly enough, it is this work that orthodox Freudians still warmly recommend to their students even today. After all, it is the book from which dream analysis can be learned. Today I consider it outdated and old-fashioned, I actually should rewrite it entirely. I had a second somewhat abridged edition appear, but I plan to write a new volume entitled *Fortschritte der Traumdeutung* [Stekel, 1935]).

I'll never forget the discussions about this book. One after the other the gentlemen pointed out to me how superficial it was. Tausk spoke for almost one hour about the linguistic errors in the preface. Freud reproached me for having used the prepositions *auf* [on] and *in* [in] erroneously three times. And so it went on. Only Adler, who sympathized with me, spoke in an appreciative way about the effort I had made.

I took it lightly, however. I knew what was coming and I had not expected anything else. I quite often saw Freud turn pale and become nervous when one of his students dared to speak his *own* thoughts. He once admitted to me in a weak moment: "I always reject everything that is new, and I need a latency period in order to overcome this aversion and to rejoice in the new." In all fairness, I must admit that Freud tacitly accepted almost everything that had been brought forward during this debate on death symbolism. Indeed, he accepted a lot in such a way that it appeared that it was his own discovery, and he avoided mentioning my name. But back to the debate. I gave a speech which I would like to preserve verbatim for posterity:

> Gentlemen! An architect once completed a great building on which he had been working for many years. He was satisfied with the delicate façade, the tastefully decorated rooms, the layout of the rooms and so on. He believed he had a right to be proud of his work, and invited his colleagues to visit it and pass their judgment. They came in large numbers, looked at the building very carefully, examined its every aspect, and kept silent until one of them noticed that the water closet didn't have a window.
>
> When everybody had seen everything, the architect asked them what they thought.
>
> "The water closet doesn't have a window," said one.
>
> "You forgot to ventilate the water closet," said another.
>
> And so it went on. They spoke for hours solely about the badly ventilated closet.
>
> "Gentlemen", the architect cried, "tell me what you think of the exquisite façade, the layout, the decoration of the rooms!"
>
> "You spoiled the water closet," a voice resounded.
>
> "But what about the new staircase, the layout of the separate elements of the house, the connection between various motifs ..."
>
> "The water closet is badly ventilated."

And so it went on and on. He listened to one critic after another, but all they mentioned was the water closet. Gentlemen! I have nothing to add to this story.

I left the room and pondered about human pettiness and revenge.

Adler saw further ahead. He always used to say: 'This is not about you. This is about me!' I thought he was wrong, because Freud surpassed himself in trying to charm Adler; he flattered him and tried to win his sympathy. Adler called him his "soul-catcher" and did not trust him. Finally, Adler was instructed to expound his hypothesis. He had at the time hit on his "masculine protest" and believed to have found in it the key to the problem of neurosis. However, the idea of masculine protest had, at the time, not yet been desexualized.

"We all want to learn!" Freud said. "Explain your theories to us."

Adler felt extremely happy and self-conscious. He planned to speak for three consecutive evenings about his doctrines. Then there was to be a discussion. One solid month of Adler's teachings! This was something different for a change in the annals of the analytical meetings, during which only Freud and Freud-related themes had been discussed.

This time Adler was not sceptical and he truly believed Freud had changed. Finally the master would accept something new, and allow his teachings to be amended and expanded. We had many discussions, yet we did not foresee the approaching storm.

Adler spoke for three evenings and developed new ideas that were later published in his book [*Über den*] *Nervösen Charakter* [On the Nervous Character, Adler 1912]. (NB: I suggested he write this book and I helped him sign a contract with Bergmann.) Then came the fourth evening of discussion. We now noticed we were facing a formal conspiracy. The first speaker was Reitler. He had prepared his response on paper and read from his manuscript. His conclusion was that what was new in Adler's theory was already known, and what was really new was not worth anything. Rosenstein spoke next in the same vein. And so it went on; one condemnation after another. I cannot remember if what now followed took place that evening or the next. Freud came to speak and he dismissed all of Adler's innovative ideas energetically and somewhat mockingly. He was visibly excited and one could see that he had prepared himself thoroughly.

Some of Adler's followers defended him, and I too paid tribute to him in a way that I deemed appropriate.

Now came the surprise.

Dr. Maximillian Steiner took the floor. Adler's teachings no longer belong to analysis. Adler is an opponent of Freud, he does not belong in the society. So that was what they were up to! Adler had to be expelled from the society!

The final decision was postponed until the next meeting. It took place in a café (Café Arkaden), in one of its medical club rooms. The atmosphere was heated. I vigorously defended Adler and the freedom of research. But I and the Adlerians were outvoted. Adler left the room with his followers. (His followers were mostly socialists, party members. They were, as far as I can remember: Furtmüller, Prof. Oppenheim, the Grüner brothers and others, about eight or ten in total.)

My first response was to join them. But I had just done something which would have complicated my withdrawal from the society. I had given up my practice as a general practitioner and had established myself as a nerve specialist in analysis. Freud had promised he would support me. I was dependent on him again, and he knew it.

I also found it difficult to give up the *Zentralblatt* which I had founded. So I stayed—and for a while Freud showered me with compliments. In one session that took place after Adler seceded, he claimed that Adler suffered from paranoia. That was one of Freud's favourite diagnoses; he had applied it to other important friends of his as well. Immediately in his slavish choir, voices resounded which enthusiastically confirmed this ridiculous diagnosis.

Freud clearly tried his trick of soul catching with me. He explained why Adler and not Stekel had been appointed chairman of the Vienna Society. He knew Stekel was not as sensitive as Adler, nor was he as ambitious. Stekel won't mind playing the second fiddle.

To me he said: "I overlooked a giant next to me and made a dwarf big. One of your symbols has more meaning for analysis than all of Adler's speculations."

He sent me a valuable agate plate from Karlsbad, which I still have, and wrote in a letter: "I don't know what could ever tear us apart ..."

But he already suspected at that time that he would one day get rid of me, just as he had got rid of Adler.

I now come to the most important event—my break with Freud. Freud writes about it in his "On the History of the Psychoanalytic Movement" (1914, p. 47): "From the third number of the second volume onwards (Winter, 1912), Stekel became solely responsible for its [Zentralblatt] contents. His behaviour, of which it is not easy to publish an account, had compelled me to resign the editorship [...]"

What does he mean by: "not easy to publish an account of"? Such mysterious allusions verge on insults and slander because anyone would understand this expression to be as if I had stolen his silverware, as Wittels aptly remarked in his book on Freud (Wittels, 1924, p. 216).

Freud clearly avoids telling the truth here, just as he did on the previous page of his aforementioned article when he wrote [Freud, 1914, p. 47] that "[On] account of scientific differences, Dr. Adler had decided to withdraw voluntarily (sic!) from the editorship."

In that same essay Freud wrote that he was extraordinarily revengeful. Unfortunately I have to maintain the opposite. I have never in my life met another man of the same intellectual stature who was endowed with the same pathological degree of vengefulness. He has persecuted and taunted me and tried to harm me in a narrow-minded way, and referred to me as someone who has gone "totally astray," which I adopted, however, as an honorary title. For it would be a good idea for the analytical doctrine if Freud could have at his command a dozen such students who had gone "astray"! Considering the enormous scientific merits that I can look back on with pleasure, I don't take it as an insult. But it is a sign of enormous ingratitude, because Freud knows very well that the majority of analysts came to Freud through Stekel. And if they don't stick with Stekel, so much the worse for analytic praxis. Freud is the ingenious theorist, I am more practical. Freud is the metapsychological philosopher, I am a realistic observer, I construct my compelling conclusions from clinical cases.

Now to some historical dates. In April 1911 discussions of my book *Die Sprache des Traumes* [Stekel, 1911c] took place. In September 1911 the third psychoanalytic congress was held in Weimar. Adler and his followers were absent. I published an anniversary issue of the *Zentralblatt* that included a portrait of Freud. During the congress there were no signs of discord or dissension. But I did not submit a presentation. Why engage in ridiculous discussions with people who only wish you harm? Bleuler spoke on his theory of autism, Jung on his contributions to symbolism, Bjerre on a cured case of paranoia, and Freud presented a "Postscript" to his analysis of Schreber's case [Freud, 1911].

Freud spoke about Schreber's delusions. He was an eagle, he said, because only eagles are allowed to look into the sun. I aroused incessant laughter when I made a few remarks about the interesting choice of topic.

Had Freud in Vienna not also had an eagle[9] who had dared look into the Freudian sun unpunished?

This remark hit the nail on the head. Freud had taken Adler's resignation too lightly. Adler became a terrible opponent. Because he left sexuality out of his doctrines and emphasized the role of aggrieved pride, he won over all those who objected to Freud's teaching on personal grounds. It is common knowledge that most people have a blind spot with regard to their sexuality. For the blind of the soul, Freud represented an eternal danger, Adler a solution.

Adler took bitter revenge for the humiliation. He chose to fight Freud, covered up the weaknesses in his system with penetrating wit, and developed a new doctrine of the power of sheer will. He renounced the title of "psychoanalyst" and called his teachings "individual psychology." In reality there is nothing individual in Adler's psychology. All mental and spiritual illnesses lead back to one and the same cause. But as a "social psychology" Adler's teachings make an important contribution to the Freudian theories and hypotheses.

Be that as it may, Freud could have learned from Adler's success how dangerous it is to push a talented pupil into opposition. However, since he, according to his own admission, badly needed an enemy, he always created one whenever he received too much recognition. Only in the roar of cannons and the smell of gunpowder does he feel good.

I already sensed in Weimar that they were preparing something against me. This was the first time that I did not give a speech, and I only got involved in the discussions. However, I enjoyed revealing the emptiness of my hidden opponents and bringing some life into the bloodless speeches this way.

In Vienna the bubble was about to burst. Freud often repeated how difficult his position was, and how he had to protect me against the Swiss. The accusations that I brought harm to analysis and made it look unreasonable did not stop. But I had an important weapon to hand: the *Zentralblatt*. I zealously wrote reviews, always objectively. However, this weapon would soon be taken from me.

The attacks by Tausk became ever more serious. I soon noticed though that they did not come from Tausk but from Freud himself, and that Tausk was a mere battering-ram, used by Freud to take stronghold "Stekel."

Remember, however, that Freud and I had the mutual right of veto. Accepting an article by Tausk was, for me, out of the question. (He had not written anything at that time, nor had he even made an attempt to write an article.) But he convinced Freud of the plan to divide the reviews among regular reviewers according to specific viewpoints. The *Jahrbuch* belonged to Jung. In the *Zentralblatt* I could criticize his works if I wanted to (I had never done so), now this possibility had to be prevented in advance. One day Freud proposed leaving all the reviews of the *Jahrbuch* as well as the

[9] In German, "Adler" means "eagle".

analytical literature to Tausk. Now Tausk had the opportunity to review my work in my own journal. I had not the slightest doubt about what form such a review would take.

But Freud was only out to intimidate me and to let me feel his power. He might just as well have used Federn, Hitschmann or any another analyst against me, against whom I could not have objected. After all, I was a far too autonomous spirit to claim the petty reviews just for myself.

I let Freud know (through Jekels) that I took this proposal as a declaration of war. I could not accept Tausk using the journal that I had founded and edited for his self-important business. I proposed to him that he nominate another reviewer if he would not let me choose one.

I told Freud he had to choose between Tausk and me. One must be aware of Freud's infinite ambition and self-will to understand his demand that I surrender unconditionally.

He reckoned that I had given up my flourishing practice and that I was dependent on his help. He trusted my accommodating nature and thought that I was weaker than I actually was.

In the meantime, Bergmann assured me that, should Freud and Stekel split up, he would trust me with the directorship of the journal. Bergmann was seriously disgruntled by the founding of *Imago* and had tried, in vain, to prevent the publication of this rival journal. We would gladly increase the available space and leave the [*Schriften zur*] *angewandten Seelenkunde* more room.

It was again on a Sunday that I paid my last visit to Freud. I uttered the prophetic words: "I know that I am being sacrificed for Jung. But the next person to abandon you will be Jung. You have turned away a loyal follower, and the one in whom you now put your trust is going to disappoint you."

Freud: "It is possible that you are right. I have been warned by X before. But I'll keep my eyes open."[10]

And that is how we parted. It was the last time that I ever spoke to Freud.

Now came the incident, the humiliation, for which Freud has never forgiven me. He wrote to Bergmann, asking him to choose between us, and Bergmann chose me. Deuticke too didn't want to hear about a *Zentralblatt*.

So Freud was forced to publish the *Internationale Zeitschrift für* [*ärztliche*] *Psychoanalyse* with Heller, a member of his society.

This is the mysterious cause of my break with Freud, of which it apparently was so difficult to give an account. It is so petty that I feel ashamed even to write about it. That is to say, the occasion is petty, but what about the motives behind it? Despotism and papistry of a researcher who himself encountered so much animosity and denial that he made the freedom of individual research subservient [to his own cause].

Freud liked to pose as the great conciliator of men, referring to the large number of followers that remained loyal to him. He recounts their names:

[10] Compare Stekel's account of this meeting in his *Autobiography* (Stekel, 1950a, p. 143).

Eitingon, Ferenczi, Rank, Jones, Brill, Sachs, Pfister, van Emden, Reik. Four of them are laymen! And what do these names mean compared with Jung, Stekel, and Adler?

The names of the most important students who dared to give voice to their own ideas are actually missing from that list. I only need refer to Professor Paul Schilder, who lectured at Vienna University on psychoanalysis, and to Pötzl, and many others. What is an Eitingon compared with a Schilder?

What names are on the list of friends that Freud has lost in the course of the past years? Had he not been ashamed, had he not felt the need to show the world that he was capable of retaining important students who remained loyal to him, he would have sent a dozen more into analytical exile.

Let us look at the friends and co-workers he has lost. First of all Dr. Josef Breuer, the discoverer of analysis, to whom Freud owes his entire career and fame. (See Freud's autobiography.) We, his students, always wondered why Freud had completely broken off relations with Breuer. Once—it was after Adler's withdrawal—I asked him about the reasons for their dissension.

Freud's account astonished me. Breuer had not followed him along the path of sexuality. At the time, *Dozent* Freud gave a lecture entitled "The Aetiology of Neuroses" in the Vienna medical *Doktorenkollegium*. After the lecture Breuer took part in the ensuing discussion. He pointed out to the society the significance of the communications. One had witnessed the birth of a new science, about which one would hear more in the future.

On his way home, Freud asked Breuer: "How come you spoke in my defence when you usually oppose my viewpoints?"

Breuer: "I wanted to help you, and to protect you!"

Freud: "So it had nothing to do with your own convictions? In that case I no longer want to have anything to do with you ..."

He turned his back on him and walked away.

And ever since, their relations have remained broken! Was ever a friend lost in such a weird way?![10] Could not Freud, if he had been more appeasing, have found his way back to Breuer later? Should he, being an analyst, not have admitted that Breuer hesitated and was affectively open to Freud because his "id" indicated him to do so? Is that a reason to avoid someone for several decades, and to refuse to reconcile with him?

I will let public opinion be the judge of this.

Now to the case of Fliess. Wilhelm Fliess, a renowned Berlin researcher, was one of Freud's most intimate friends. Their friendship was so important that, for example, they took Easter holidays together in Prague. (Fliess

[10] Allusion to Shakespeare, *King Richard III* (I, ii, 227–228): "Was ever woman in this humour woo'd? Was ever woman in this humour won?" The original reads: "War je in solcher laun' ein Freund verloren?!"

worked in Berlin, Freud in Vienna.) It was Fliess who drew Freud's attention to the law-like principle of bisexuality. Freud himself told us of this in our analytical society even before he broke with Fliess.

Then came the distasteful plagiarism incident which involved Fliess, Weininger, and Freud. Fliess reproached Freud for having told Weininger the law of bisexuality and Swoboda the principle of periodicity.

Weininger and Swoboda have voices of their own, they have no need to borrow other people's ideas. I do not wish to establish whether or not a case of plagiarism occurred. I have repeatedly said it was not the case. The idea of bisexuality was known long before Fliess, and Swoboda discovered and provided proof for his theory of periodicity independently of Fliess, and at any rate quoted from him and gave references to his work.

Here was another friend that Freud lost. Of course here too he gave the diagnosis he always had at the ready: paranoia. A clear case of conspiracy delusions, therefore. Would anyone else sacrifice so easily the only loyal and kindred spirited friend that he had?

Now I must speak of the desertion of Jung and Bleuler. I did not witness it personally and do not wish to appeal to questionable witnesses. Jung's *Wandlungen* [*und Symbole*] *der Libido* [Transformation of the Libido, 1911–1912] was too ambitious in design to be accepted by the master without opposition. The concept of libido was transformed in Jung's hands into a force that could just as well be called energy instead of libido. But Jung was extremely widely-read and had considerable ethnographic expertise. So when Freud came to speak about "totem and taboo" [Freud, 1912–1913] in our society, I immediately realized that the Freud–Jung duel had begun. The rift came at the Munich congress. Vienna and Zurich flew at each other's throats. More about these scientific differences are to be found in Freud's account. Personally, I do not believe in these so-called scientific differences, and believe them to be personal. Jung was tired of always having to play Freud's crown prince, and more and more rivals also came to the fore.

Whatever happened between Bleuler and Freud happened without my knowing. The fact is that these two brilliant men had drifted so far apart while they could have mutually and fruitfully influenced each other.

I do not want to conceal that one of our oldest members, the congenial Max Kahane, also came into conflict with Freud. I have never asked Kahane the reason why, but the way he came to speak about Freud cannot possibly be reproduced here. Not that he ever doubted his scientific merits; it was the way Freud treated his friends, which is what Kahane once considered himself to be.

I believe this list suffices. It could perhaps be added to. Many students of Freud were lucky enough to live far away from the master, and they were able to deal with his sensitivities. They would have found it easy. Breaches in the association may already have occurred, students may have already fallen from his grace.

I therefore find myself in good company. I am not an exception. Or perhaps I am, considering the irreconcilable hate Freud felt for me. I have already mentioned his scorn for me in his works. And on top of that, there was the order that I was not to be quoted! *Anathema sit!* [Let him be accursed.] Everything I discovered was considered common property or was attributed to Freud. I can give countless examples of that, but I do not want to get bogged down in petty details. Somewhere Freud permits himself the following expression: "Reluctant as I am, I must quote Stekel here."[11] Analysts who kept company with me were excommunicated or fell from Freud's grace. (An example of this is the brilliant Herbert Silberer.[f])

When I tried to set up in New York a new analytical periodical "Psyche and Eros" with Tannenbaum, Freud wrote to Tannenbaum that I was "a terrible man,"[12] and he referred to me as one of his worst enemies.

Now let me ask one question: What makes Freud say I am one of his worst enemies? Has an analyst ever done so much for analysis as I have? Did anyone disseminate as much propaganda for analysis as I did? Has anyone devoted the same amount of energy to trying to prove the importance and fruitfulness of analysis? Did anyone else have the courage to come to the fore with his clinical work and say: we owe all this to Freud? Show me the sentences in my *Disorders of the Instincts and the Emotions* series where I look down on Freud. I contradict him, fight him, but I also acknowledge the great importance of his work. I can separate Freud the human being from Freud the scientist. It is Freud's tragedy that he has never been analysed himself. He needed it just as much as many of his students.

The greatness of a man is not only expressed in his intellectual talents, it also shows in his ability to admit his errors to himself, to improve himself, and to be able forgive and ... to forget. But Freud cannot forget. He resembles his god, the furious Jehovah, who had Korah and his gang engulfed by the earth, and who plunged his most beloved son Job into misfortune solely to put his faithfulness to the test. But not everybody is born to suffer and to praise god, like Job, and the analytical gang of Korah only fell into the depths of the earth to rise again even higher than before.

[11] Compare the following passage from "Dreams and Telepathy" (Freud, 1922): "I once dreamt during the war that one of my sons then serving at the front had been killed. This was not directly stated in the dream, but was expressed in an unmistakable manner, by means of the well-known death-symbolism of which an account was first given by Stekel. (We must not omit to fulfil the duty, often felt to be inconvenient, of making literary acknowledgements.)" The phrase "Reluctantly, I must quote Stekel," does not seem to appear anywhere in Freud's works, but perhaps Stekel had in mind an addition to the 6th edition of *The Interpretation of Dreams* (Freud, 1900, p. 350), in which Freud describes how "it was only by degree and as my experience increased that I arrived at a full appreciation of its [symbols in dreams] extent and significance, and I did so under the influence of the contribution of Wilhelm Stekel."

[12] "terrible man" in English in the original.

I do not want to compare myself with Freud. I never have. But I want justice. I received a great deal from Freud. But I gave at least as much, if not more, in return. We could call it quits. And at any rate, my separation didn't do the analytical society any good. Just look at the last publications: metaphysical and meta-psychological speculations, endless fantasies about castration complexes; compare my *Disorders* series with that. That is not speculation, that is clinical work.

I demand grave objectivity. I will not let my work, to which I have dedicated my life, be distracted because of an envious person. I need only refer to the problem of the ego ideal. I have, for many years, pointed to the moral tendencies in mankind, for years Silberer had spoken of anagogical tendencies. In 1913 my work on fetishism appeared, in which I related the religious part of the ego to repressed morality. In my book *Die Geschlechtskälte der Frau* [Frigidity in Woman, Stekel, 1921b], and in *Die Impotenz des Mannes* [Impotence in the Male, Stekel, 1920] (*Disorders* series, Vol. 4) I pointed out how man and his actions are determined by the secret ethical ideals. Then we read in Freud in his book *The Ego and the Id* [1923, p. 12]:

> In these pages things are touched on which have not yet been the subject of psychoanalytic consideration, and it has not been possible to avoid trenching upon some theories which have been put forward by nonanalysts or by former analysts on their retreat from analysis. I have elsewhere always been ready to acknowledge what I owe to other workers; but in this instance I feel burdened by no such debt of gratitude. If psychoanalysis has not hitherto shown its appreciation of certain things, this has never been because it overlooked their achievement or sought to deny their importance, but because it followed a particular path, which had not yet led so far. And finally, when it has reached them, things have a different look to it from what they have to others.

That may sound like a proud statement, but it does not correspond with scientific custom.[13] If I were to have learned something from my bitterest enemy, I would quote him. Freud is always so sensitive about not being referred to. It is preposterous how analytical works are always honouring Freud. "Freud said"—it already sounds like a bible quote from a preacher. I recall a little episode when I published a paper in the *Berliner Zeitgeist* on "superior ideas" and I quoted Wernicke. Freud read the article with great interest and said at the end a little maliciously: "You thought you could forget me!"

How often could I have said that to Freud and his followers. But I hate petty priority struggles. Ideas are in the air. I am convinced that the world would have known another Freud if this one had not been born. Nietzsche and Schopenhauer said everything that Freud later also said. But has it

[13] This sentence strongly resembles a line in Freud's letter to Stekel dd. 13.1.1924, when he denies that he broke with him on account of scientific differences: "This sounds quite good in public, but it doesn't correspond to the truth."

occurred to anyone to accuse Freud of plagiarism? No. But remember that Nietzsche wrote the unforgettable words: "You serve your master wrongly if you don't transcend him."[14] I have never been so arrogant as to claim that I am of the same intellectual stature as Freud. I have used the comparison that a dwarf standing on the shoulders of a giant can see further than the giant. That would diminish my claim somewhat.

Given my own experiences, I cannot recommend the company of a giant to anyone. They are cannibals, and not only in fairytales. The influence of a genius is best experienced at a distance, when you don't have to look and laugh at his petty weaknesses. Freud wrote a nice essay on "The Antithetical Meaning of Primal Words" [Freud, 1910], in which he says that there was a time when "big" meant "small," when the same expression was used for a giant and a dwarf. Does that not mean that every giant is also a dwarf, and every dwarf a giant? *Ach*—it all depends on how you look at it. I would have looked at Freud forever only with the eyes of love had he not forced me to remember The master, who discovered repression, should, of all people, know that "repressed students" can return as conversion symptoms. They become converted. But not me. I flatter myself with the idea that I am the only true Freudian. I now represent Freud much better than he does himself. That is what I hope people will say about me one day. But today I can already say that I have the majority of medical doctors at my side. The society of independent medical analysts, only 2 years old, already has more doctors than the old Freudian organization. Of the 41 members of the Vienna analytical society, 13 (!) are lay analysts!

In his *Autobiography*, Freud [1925, p. 70] says about lay analysis:

> It is no longer possible to restrict the practise of psychoanalysis to doctors and to exclude laymen from it. In fact, a doctor who has not been through a special training is, in spite of his diploma, a layman in analysis, and a nondoctor who has been suitably trained, can carry out the analytic treatment not only of children but also of neurotics.

I really must object to these statements. A medical doctor who practices analysis is not half as dangerous as a layman who practices analysis without proper medical training. The layman can easily overlook a life threatening symptom. Take the case of a patient with an intestine neurosis who complains of physical pain. This pain may be the first symptom of an appendicitis. The layman does not recognize the symptoms and it may be too late for a life-saving operation! And there are a thousands of possibilities like this. However, if the layman sends the patients to a doctor, then the doctor can check whether or not it is a parasympathic symptom. But the layman

[14] Allusion to *Thus Spoke Zarathustra*: "Now I order you to lose me and to find yourselves, and only when you have rejected me will I return to you." Jung too referred to that same sentence in his letter to Freud dated 3 March 1912.

loses his authority. Irregular supervision by a medical doctor is absurd and inexpedient. Collaboration with a medical doctor, who checks the patient regularly and receives daily reports, however, is a possibility, when the analyst is a healing assistant.

But Freud speaks *pro domo* [for the home]. As his legacy, he wants to pass on analysis to enable his best students, including the lay analysts Rank, Reich, Sachs, Aichhorn, Lou Andreas-Salomé, Bernfeld, and perhaps his daughter, Anna Freud, to make a living.

I know from my own experience that a thorough training in internal medicine and neurology signifies a successful analysis. Experiments with autonomous lay analysts (Herbert Silberer, who was surely a gifted analyst) resulted in failure. Freud knows very well that he would not have arrived at his analytical discoveries without medical knowledge.

These words of Freud will be idolized in the further development of psychoanalysis. Even today there are many bad cases of parapathia [neurosis] who have been through analysis and work as analysts. Every tormented soul wants to make a living as an analyst. Analysis gives the analyst a feeling of power; transference, otherwise one of the most powerful weapons in the healing process, can also send patients into dependency on swindlers and charlatans.

Freud should have known all this. Or was that sentence addressed to me, because I set up the "organization of independent medical analysts"? Or does he simply want to hand over his material legacy to his students? I do not believe that Freud is driven by such low motives. I believe it is an expression of gratitude to his nonmedical friends, for having remained loyal to him.

Freud is supposed to have said: "My best students are laymen." But who can guarantee that charlatans will not abuse analysis by referring to Freud? Analysis without proper medical knowledge is wrong. Symptoms may turn up every day which only a medical doctor can decide whether they are organically or psychologically based. One should not degrade analysis to a form of Couéism. Would that be the last victory of Freud over me? Once more a ridiculous game: "Who will succeed me? Who will acquire my inheritance?"

Science has no legacies. You don't appoint an analytical successor. Or you do, and the world brings him down the next day.

Analysis may be a living for many analysts. For me it is an artistic occupation. I want to heal patients and to educate students. And I enjoy students who transcend me, from whom I can learn. No-one is infallible, least of all he who believes in his own infallibility.

I do not want to conceal that I repeatedly attempted a reconciliation with Freud. I said before that I am good natured and forgiving by nature. I can forgive and forget. Freud cannot. I wrote to him on the occasion of an insignificant event, and I said that I would never forget that it was he who lit the torch with which I found my way. I hoped he would acknowledge his errors, and I tried to elicit from him a statement to that effect. In vain!

The objective of this essay is not to detract from the merits of a brilliant man who expanded the limits of our knowledge almost infinitely far. *But where there is much light, there is also much shadow.*

I bid my friends to see in this essay nothing more than a kind of extorted self-defence. The shadows thrown by this genius threaten to disturb my light. I owe it to my self to expound the facts as they occurred. It hurt me that I have been forced to do so. I have struggled for years with it, and finally found myself forced to do it because the offensive account of the master is too important to simply ignore it and continue as if nothing had happened. Freud has had ample opportunity to backtrack and correct himself. His respect for my work could have made him do that. But no, he and his students outdo each other in their ridiculing and deprecating of my work. Read the critiques of my work in orthodox analytic periodicals and you will understand my position. After all, no one likes to be represented as a loser in public. I don't claim to be a genius. Perhaps that is why I am uninhibited. I attempt to acknowledge my mistake, and I'm not ashamed to admit when I'm wrong. In spite of everything that Freud has done to me, I am still grateful to him, and hold him in higher esteem than many of his followers who claim to be his best friend. What I have done for him, and still do, I have done with pleasure and without envy. I gladly bow to such an outstanding genius, to whom the words of the poet [C.F. Meyer] apply:

> I am not a book that's filled with clever fiction
> I am a human being, in all my contradiction

Vienna, June, 1925.

Notes

a. The section reads (Freud, 1896, p. 207): "It seems to me certain that our children are far more often exposed to sexual assaults than the few precautions taken by parents in this connection would lead us to expect. When I first made enquiries about what was known on the subject, I learnt from colleagues that there are several publications by paediatricians which stigmatise the frequency of sexual practices by nurses and nursery maids, carried out even on infants in arms; and in the last few weeks I have come across a discussion of 'Coitus in Childhood' by Dr. Stekel (1895) in Vienna."

b. In my own school, the younger students must report to me daily; the older ones report their cases in weekly meetings. Mistakes are corrected and collectively discussed.

c. Case no. 108, Chapter 18 "analysis of a heart neurosis," Vol. I, 4th edition, p. 287.

d. Emphasis in the original.
e. This claim is repeated to this day by many Freudians. I say, I would be proud if it were true. Because then I would equal a poet whose creativity and fantasy would not and could not be exceeded by anyone. My pupils, to whom I present various cases every week, and who I provided with material, know how large my experience is. Taking into account all students and assistants, there are on average between 60 and 100 patients under my supervision. Everybody profits from this material because the most interesting cases are presented and controlled on a weekly basis.
f. See my obituary for Silberer in the first Volume of the *Fortschritte der Sexualwissenschaft und Psychoanalyse.*

Chapter 9

THE STEKEL–FREUD CORRESPONDENCE

Edited and translated by Jaap Bos and Paul Roazen

A note on the transcription of the correspondence: all 5 of Freud's letters (items 1, 2, 8, 27 and 54) are handwritten. Freud's letter to Stekel dated 13.1.1924 was previously published in *The Letters of Sigmund Freud* (E. Freud, ed., 1960). The first five letters by Stekel are also handwritten (items 3–7), all others are typed on either his own stationery or on that of the *Zentralblatt* (to 1912, after which various stationery is used). Most, but not all of Stekel's letters are signed. Minor misspellings (such as Datner for Dattner, Necken for Nelken, Lutzenberger and Luzzenberger for Luzenberger, etc.) and typing errors were tacitly corrected. Abbreviations (Z.B. for *Zentralblatt*, etc.) have been spelled out. Uncertain readings and completions are placed between brackets. Freud's letters were transcribed by Ernst Falzeder, whom we kindly thank. We also thank Dave Lee for his assistance in translating the letters into English. This correspondence also contains 3 additional notes by Stekel to a certain Herr Rat and a letter by Stekel to Jung (the relevance of which is discussed in chapter 7). All items were taken from the Library of Congress Freud collection. Freud's letters are published with permission of Sigmund Freud Copyrights/Mark Paterson Marsh.

[1] Freud to Stekel
PROF. DR. FREUD
IX. Bergasse 19
Vienna, 4.2.1904

Dear colleague
The psychological Society you founded commissioned me, after I presented them with your letter of resignation, to invite you to our midst next Wednesday "as if nothing had happened". Should you find it difficult to do

so, and wish to confer with me beforehand, I would be very pleased to assent to such a proposal of yours.

With amicable greetings
Freud

[2] Freud to Stekel
Vienna II, Castellezgasse 2
Aussee 16. 8. 05 [postcard]

Dear colleague
Received with thanks and read it with great interest, I was greatly annoyed by the "social" misprint, then saw the corrections.[1] I hope it will greatly promote the circulation of the book.

Cordial regards
Freud

[3] Stekel to Freud
MED. DR. WILHELM STEKEL
I. Bez. Gonzagagasse 21
Vienna, 28.2.1910

Dear Professor!
Jones' idea is excellent.[2] I will gladly present a talk at the conference, illustrated with a few examples about the practical work; I shall write to Zurich today.

I still have not yet fully recuperated from "shock".[3] I have first of all the impression that I said something really important, and secondly, that it was one of the guests who made fun of me and treated me as a sampler. It hurt me most, however, that you did not say anything about tolerance to Sadger. I am otherwise not sensitive and I accept much of what I am told. Also I realize that jealousy has a malicious tone. In short, it was more than I could bear.

But enough of my own petty injuries. I am sending you my presentation by way of substitute. It is entirely didactical and can serve as propaganda.

[1] Reference unknown.

[2] In pencil in a different handwriting "Sammelkommittee Symbolik" [assembling committee on symbolism]. At the second psychoanalytic congress, 30–31 March 1910, at Nuremberg, Stekel proposed the formation of a committee to collect reports on symbolism in psychoanalysis for publication. The proposal originally came from Jones (see Jones to Freud, 12.12.1910).

[3] "Shock" refers to the strong critique which Stekel's paper on the psychology of doubt met on 19.1.1910 at the Vienna Society. Stekel complained that "everything new that he offers encounters great resistance at first, only to be accepted in many instances after publication" (*Minutes*, Vol. II, p. 403). The "guest" who mocked Stekel was Binswanger, who was present at this meeting and criticized Stekel's paper for blurring many things together. Sadger is recorded to have said that Stekel's presentation was a mere juggling with words (*Minutes*, Vol. II, p. 399).

I will let the work on doubt mature a little further. From it will come, I hope, a second book on anxiety.

With many thanks for your stimulation and cordial greetings,
Your grateful student
Stekel

[4] Stekel to Freud
MED. DR. WILHELM STEKEL
I. Bez. Gonzagagasse 21
Vienna, 12.4.1910

Dear Professor!
As I am still recovering, regrettably I cannot come to today's meeting. A case of retropharyagitis played up badly as well.[4]

I spoke with Deuticke[5] today on the telephone. He cannot and does not want to decide anything without consulting you. I believe that a resolution is therefore in your hand. With psychoanalysis all is well at the moment, despite *July* cases. Some came to an end, one took refuge in flight, one began a grand tour. I suppose every analyst has to deal with this. I shall leave for a unexpected holiday. At least illness did not cause too much damage.

I am already looking forward to the work the new monthly will bring us.[6] All Viennese must be put to work. Including a certain Dr. Stekel, who greets you cordially.

P.S. Kraus attack leaves little to guess about his evil intentions.[7] Is that only foreplay, -but who is going to [illegible]?![8]

Please [beg] the gentleman humble forgiveness.

Yours affectionately
Stekel

[5] Stekel to Freud
MED. DR. WILHELM STEKEL
I. Bez. Gonzagagasse 21
Vienna, 24.4.1910

[4] Retropharyngitis: inflammation of the posterior part of the pharynx, causing acute neck pain.

[5] Franz Deuticke, Viennese publisher of most of Freud's early works.

[6] The *Zentralblatt* was officially founded at the third psychoanalytic congress in March 1910 and the first issue appeared in September of that year. Bergmann, who also published Stekel, undertook the publication of the journal.

[7] Presumably a reference to Karl Kraus, *Die Fackel* of 9.4.1910, which carried several ironic and sarcastic attacks against psychoanalysis (see Szasz, 1977, pp. 103–104), from which we quote the following aphorism: "Psychoanalysis is the occupation of lewd and lascivious rationalists who attribute everything in the world, except what they themselves do, to repressed sexuality". Timms (1986, p. 110) argued that Stekel, not Freud, was the main target of Kraus' later attacks on psychoanalysis.

[8] Reading of the second part of the sentence is uncertain: it perhaps contains an allusion to musical fugues.

Dear Professor!
The latest issue of the *Deutsche Medizinische Wochenschrift* carries an obtuse paper by Romberg on Hysteria.[9] Hysteria is [cured] by way of crudest association work!! The best for the last. The Freudian method is unappropriate for "*German* women". Meaning: it is a specifically Jewish treatment. Must one only leave something like that unanswered? I could show this gentleman some primal German female specimens ... .

Thanks very much for the dream symbols. Both examples are very [top of page missing]. I have some very beautiful examples of biographical sexual-biographical dreams.

Deuticke already called off even before I got to speak to him personally. I explained to him that these publications will also be bought by the lay public.

Obviously we will start a strictly scientific journal, but that should not make it boring.

Today I met with Adler. We shall edit the pamphlet together and publish it as "Vienna Discussions" Vol. I, "On suicide in children".[10]

The Rombergs and Isserlins will become very petty.[11]

"Causes of nervousness" has already appeared in *Danish*.[12]

Best wishes
Yours gratefully
Stekel

[6] Stekel to Freud
MED. DR. WILHELM STEKEL
I. Bez. Gonzagagasse 21
Vienna, 26.4.1910

Dear Professor!
We have no intentions at all to undermine your institutions. We are too much interwoven with you and the cause as to not to want anything else but to reach the highest level. There will be absolutely no endless polemics! Only the barest essentials *allowed* or *requested by you*.

We want to discuss further details in a café tomorrow. We shall meet your wishes in any way we can. After all, that is in our interest as well.

[9] Romberg (1910). Ueber Wesen und Behandlung der Hysteria. *Deutsche Medizinische Wochenschrift*, *16*, pp. 737–742.

[10] Adler et al. (1910). *Über den Selbstmord insbesondere den Schüler-Selbstmord. Diskussionen des Wiener psychoanalytischen Vereins*. Wiesbaden: J. F. Bergmann.

[11] Max Isserlin (1879–1941), assistant of Kraepelin. The reference is probably to Isserlin (1910). Die Psychoanalytische Methode Freuds. *Zeitschrift für die gesamte Neurologie und Psychiatrie*. See letter of Freud to Ferenczi, dated 12.4.1910 and note.

[12] Stekel (1907a). *Die Ursachen der Nervosität*. Vienna: Bergmann.

With many thanks and cordial greetings
Your loyal
Stekel

[7] Stekel to Freud
MED. DR. WILHELM STEKEL
I. Bez. Gonzagagasse 21
Vienna, 24.5.1910

Dear Herr Professor!
Many thanks for the contribution.[13] It is enormously important ...

I must to press Bergmann very hard for exchange copies.

I think we should sacrifice the two redundant copies to *both*.

I shall give one to the Society, send it to [Zurich], Adler, etc.

Once the deal starts to come off, Bergmann will be pleased to make further copies available.[14]

This is naturally only my private opinion. If you wish I shall write to Bergmann ...

Today I will finish editing the issue. I fear there two. I claimed 4 proofs from Bergmann.

Poetry will be cut out entirely (with the exception of the poem by Nietzsche).

Best wishes and greetings,
Yours affectionately
Stekel
P.S. Today I made an important numerological discovery regarding the ... Ten Commandments! Please take heed!

[8] Freud to Stekel
18.6.1910 [Postcard]

Dear Doctor
I would like to propose sending Herr Dozent Dr. M. Brahn in Leipzig, Waldstrasse 56, publisher of the *Zeitschrift für paedagogische Psychologie* the announcement of the *Zentralblatt*. He is truly sympathetic towards psychoanalysis.

Cordially
Yours
Freud

[13] Presumably Freud (1910d). 'The Future Prospects of Psychoanalytic Therapy', first published in *Zentralblatt*, I (1), pp. 11–19, but this is uncertain since Freud wrote to Jung only on 5 July 1910 that he had furnished this paper.

[14] On 17 May 1910 Freud wrote to Jung that the contract with Bergmann was "already signed" and that the conditions were "most favourable".

[9] Stekel to Freud
ZENTRALBLATT FÜR PSYCHOANALYSE
Vienna, 14.12.1910

Dear Professor!
Yesterday I was somewhat confused and left quietly.[15] Unfortunately, I am quite at a loss when philosophers start to talk. I can't help it. I surely must have a blind spot for philosophy.

I am sending you two important letters. Luzenberger is really naïve. He doesn't read the *Jahrbuch* and he wants to work for the *Zentralblatt*.[16]

I was right about Assagioli.[17] I shall send the papers for review to Maeder.[18] He will give him a reply.

Urgently request from you the address of *Storfer* in Zurich.[19]

With cordial greetings,
Yours loyally
Stekel

[10] Stekel to Freud
MED. DR. WILHELM STEKEL
I. Bez. Gonzagagasse 21
Vienna, 2.4.1911

My dear Professor!
How could you believe that I wanted to rob you of the right to open critique![20] We are engaged in science. Do I demand respect or even praise from you. No! Do write to me and tell me what you think. I have actually been treated in recent times only "monopolary." Bipolarity of opinion would already mean progress to me.

Today I wrote to Bergmann to remind him to send you the proofs punctually. The next issue is still chaos for the time being. As far as I can see, it

[15] Letter may be misdated. The meeting of the Vienna Society took place on December 14. Stekel's presence is recorded in the attendance list but no comments of his appear in the minutes. The discussions of that evening were dedicated to "brief clinical as well as other communications," and contains many different topics, although no particularly philosophical ones.

[16] August von Luzenberger, early member of the Swiss Psychoanalytic Society, was listed as member of the editorial board of the *Zentralblatt*. No further details known.

[17] Roberto Greco Assagioli (1888–1975), also member of the Swiss Psychoanalytic Society and likewise member of the editorial board of the *Zentralblatt*. He was a co-worker of Jung at this time and introduced to psychoanalysis by Ernest Jones. He published a few abstracts in the *Zentralblatt* of papers that he delivered elsewhere.

[18] Alphonse Maeder (1882–1971), member of the Swiss Psychoanalytic Society and editor of the *Jahrbuch*.

[19] Adolf Storfer (1888–1944), member of the Vienna Psychoanalytic Society. He played an important role in the editing of psychoanalytic journals.

[20] A reference to the conflict with Adler, in which Stekel was involved as well (see chapter 7).

will be excellent. There are some disagreeable irregularities with the printers. By the way, I need to talk to you next Wednesday about the journal a bit. Unfortunately, there will be a double issue again. Or should we split the material. More when we meet. Perhaps after the next meeting.

With cordial greetings
Yours affectionately
Stekel

[11] Stekel to Freud
MED. DR. WILHELM STEKEL
I. Bez. Gonzagagasse 21
Vienna, 9.4.1911

Dear Professor!
You are right. Right now I cannot find the quotations in Riklin.[21] I believe they are somewhere in Jung or Maeder. I shall ask about them and let you know. The reference to snake in the grass, by the way, already points to the femininity of this symbol. Also, according to Kleinpaul, the snake is a symbol for rejuvenation and renaissance.[22]

More about it face-to-face. Cordial greetings
Yours affectionately
Dr. Stekel

[12] Stekel to Freud
ZENTRALBLATT FÜR PSYCHOANALYSE
Vienna, 29.6.1911

Dear master!
I felt the need to wish you well before I leave, and to express my delight about the fact that I can love my teacher again. I hope that from now on we shall always fight shoulder to shoulder and work together for you, for our cause. I leave for Ischl tonight. My address is: *Kaltenbachstrasse 26*. Contributions for the celebratory issue keep coming in.[23] I expect we shall publish a fine issue. Salutations will reach you in Karlsbad or in Vienna still. Please also send me your address.

[21] Reference unknown.

[22] Rudolf Kleinpaul (1885–1918), philologist, author of books on the symbolical meaning of words. Kleinpaul was often quoted by the members of the Vienna Psychoanalytic Society, and Stekel in particular (Freud also devoted some attention to him in *The Interpretation of Dreams*).

[23] The last issue of the first volume of the *Zentralblatt*, published in Sept. 1911, celebrated the third psychoanalytic congress of 21–22 September 1911. It contained contributions by Putnam, Abraham, Dattner, Drosnes, Ferenczi, Heller, Hitschmann, Jones, Jung, Maeder, Marcinowski, Rank, Reitler, Robitsek, Rosenstein, Sachs, Sadger, Stekel, and Wagner.

Scripture is a European and at the moment in London. I received a very appreciative letter from him about my dream book. He wants to look me up in Ischl. I want to try to persuade him to come to the congress.[24] Should I ask him for a contribution for the celebratory issue?

Once again my cordial greetings and many wishes for an agreeable summer.
Yours loyally
Stekel

[13] Stekel to Freud
Zentralblatt für Psychoanalyse
Ischl, 2.7.1911

Enclosed two letters from Bergmann, from which you can see that he is of the same opinion about the new journal as I.[25] It will first of all hurt the *Zentralblatt*. I bid you to postpone the plans for a year, until the *Zentralblatt* has consolidated. The downfall of one or our journals (and the journal published by Heller will not last[26]) is going to produce a cry of triumph from our opponents.

With many cordial greetings
Yours loyally
Stekel

[14] Stekel to Freud
Zentralblatt für Psychoanalyse
Ischl, 12.7.1911

Dear Professor!
I have arranged everything according to your wishes. Please send enclosed proofs with your signature to Bergmann. That finishes the business.

I am recovering well. You would not recognize me. Hopefully, you are doing as well.

In haste, with many cordial greetings
Yours loyally
Dr. Stekel

[24] E.W. Scripture (1864–1945), experimental psychologist of Leipzig, later director of a research clinic at Vanderbilt Clinic, New York, and, according to Jones, highly critical of psychoanalysis. It is interesting to note what Jones wrote to Freud on 29.11.1913 about Scripture: "he helps the psychoanalytic cause in about the same way that Stekel does".

[25] The new journal was *Imago*, published in 1912 and edited by Rank and Ferenczi.

[26] Heller's journal was the *Korrespondenzblatt* (newsletter), published as an independent journal; it was merged with the *Zentralblatt* a few months later. Unknown to Stekel was that Freud had already proposed such a plan to Jung earlier, but apparently Bergmann, the publisher, was not in favour of it (see Freud to Jung, 1 March 1911).

[15] Stekel to Freud
ZENTRALBLATT FÜR PSYCHOANALYSE
Ischl, 14.7.1911

My dear Professor!
For a number of reasons, I shall not bring the last short piece of miscellaneous entitled "Flight from Women". I want to announce the third edition of *The Interpretation of Dreams* and also include the ridiculous critique from the *Zeitschrift für Strafrechtswissenschaft*.[27] It is so amusing. Why should we not permit our readers a few seconds of merriment?

Adler demanded acknowledgement again. I did everything to clear out everything for once. We don't need to be afraid of his journal any more than his Society. We just need to work hard and keep our forces together. In the autumn we shall enter into new discussions, which will prove for all to see what *we* have accomplished, and with which I mean to put you in the spotlight.

Bergmann wants permission to print at Franzos again. Endless scribblings. I am now done. The next issue will include the declaration of our independence.[28]

I hope the little digs of male protest won't disturb your cure. I am doing extremely well. They say I look ten years younger. Psychoanalysis in the woods proves itself.

With many cordial greetings
Yours loyally
Dr. Stekel

[16] Stekel to Freud
ZENTRALBLATT FÜR PSYCHOANALYSE
Ischl, 19.7.1911

My dear Herr Professor!
What pleases us most in gifts is the fact that the giver thought about us and wanted to please us.[29] You too found time during your holiday to think of me and to do me a big favour. Saxa loquuntur! [The stones speak[30]]. Accept my warmest thanks for that. Now allow me to add another debt to everything immaterial that you've already given me, which I cannot, like the previous ones, pay back except by working on our joint project, which is after all your project.

[27] *Zentralblatt*, Vol. I, nr. 10–11 (Oct–Nov. 1912), p. 530: "Die 'Traumdeutung' von Freud" and "Die Zeitschrift für Sexualwissenschaft (XXXI. Heft 7)."

[28] Issue nr. 10/11 (July/August) of the *Zentralblatt* opened with a "statement" of Adler, explaining to the reader that he retreated from the journal as editor because "Freud is of the opinion that scientific differences between him and me [Adler] prohibit joint publication of this journal".

[29] In his historic account, Stekel mentions how Freud sent him from Karlsbad a valuable plate, writing: "I do not know what could ever tear us apart ... " (see chapter 8).

[30] Possibly an allusion to *The Aetiology of Hysteria* (Freud, 1896, p. 192).

I am sending you first of all the draft of the salutation for the congress.[31] I did my best to make it sound worthy and affectionate. Perhaps the summer heat had a paralysing effect though.

The omitted spaces contain spots for top and bottom.

With many warm greetings, and many thanks, again, for the dish of love (as a symbolic replacement of the dish of wrath which you once vented upon my head).[32]
Yours loyally
Stekel

[17] Stekel to Freud
ZENTRALBLATT FÜR PSYCHOANALYSE
Ischl, [?] July 1911

Dear Herr Professor!
The material for the *Zentralblatt* pours in so lavishly that several issues are covered now. Shall we open the second volume with a very good didactic article by Wulff on "infantile sexuality"?[33]

Additional submissions are: reviews by Jones, whose letter I enclose.[34] I have already answered it. Furthermore "Jungfrau und Dirne" [virgin and prostitute] from Storfer in Zurich (on veil symbolism),[35] a case of obsessive-compulsive neurosis from Luzenberger,[36] a paper on lies in psychoanalysis from Wulff,[37] and several smaller contributions for the celebration issue. Which papers do you wish to read? We must now plan a careful selection of the contributions.

I am doing excellently here. Think of the coincidence: your dear mother[38] and sister[39] live right across the street and hear me play the piano every day. Work is really a side issue now, and psychoanalysis is reduced to quiet atonement two hours a day, and these even while walking in the woods. That is my newest technique.

I hope you recuperate fully from your heavy work.

[31] The last issue of the *Zentralblatt* contained a "salutation" (Festgruss) to the third psychoanalytic congress (in Weimar).

[32] "Wrath": a reference to the difficulties in their relationship of March–April following Adler's resignation (see the introduction to these letters).

[33] Wulff (1912). Beiträge zur infantilen Sexualität. *Zentralblatt, II* (10); pp. 6–17.

[34] *Zentralblatt*, Vol. II nr. 3 (Dec. 1911), pp. 46–50 contains 9 short reviews by Jones of Freud, Jung, Ferenczi, and a number of non-analytic authors.

[35] Storfer (1912). Jungfrau und Dirne. *Zentralblatt, II* (4), pp. 200–204.

[36] Not published.

[37] Wulff (1912). Die Lüge in der Psychoanalyse. *Zentralblatt, II* (3), pp. 130–133.

[38] Amalia Freud, née Nathanson(1835–1930). She spent many holidays at Ischl.

[39] Presumably Adolphine Freud (1862–1942).

With many cordial greetings
Yours loyally
[Stekel]
I received a formal letter from Adler. He demands several endorsements and so on. The letter is so reserved that even formal salutations at the end are missing.

[18] Stekel to Freud
ZENTRALBLATT FÜR PSYCHOANALYSE
Ischl, 7.8.1911

Dear Herr Professor!
Wulff gave me permission to carry through improvements and cutbacks as I see fit. I intend to use it in accordance with your intentions. I received very kind letters from Jones and Brill, who expressed their satisfaction that I am not part of the secession and who promise diligent cooperation. A short contribution by Brill shall go to Federn for translation.[40] The first proofs of the celebration issue should reach you any day now. Please read through the "Festgruss" [celebratory greetings] once more carefully and let me know any changes you might want.

Issue no. I of the second volume shall thus contain the following:[41]

Original contributions:	1) Nelken "On Wordbuilding with Schizophrenics"
	2) Wulff ["Contributions to Infantile Sexuality"]
	3) Stekel, "Various Forms of Transference"
Communications	1) Brill
	2) Ferenczi
	3) Jones
	4) Maeder
	5) Stekel

Reviews, Misellaneous, *********, in short, it will be a fine issue. Do you not have something small? Should we not give our readers a little introduction and underline that we emphasize the didactic aspect, or should we do that in a concluding note at the end of issue XII?

I don't hear anything from Vienna. Oppenheim was in Ischl, he saw me and he looked away.[42] Apparently I am supposed to be an evil spirit.

I wish to speak to you before Weimar because I'll be travelling to Wiesbaden beforehand.

With cordial greetings.
Yours loyally
Stekel

[40] Rank (1912). Review of A.A. Brill, Freud's theory of will. *Zentralblatt, II* (4), p. 220.

[41] The published edition has original contributions of Nelken, Wulff and Dattner (see letter 20); Stekel's original contribution became a "communication"; the second piece by Stekel was replaced with a communication by Harnik.

[42] David Ernst Oppenheim (1881–1943), member of the Vienna Psychoanalytic Society, follower of Adler; he left the society in 1911.

[19] Stekel to Freud
Zentralblatt für Psychoanalyse
Ischl, 23.8.1911

Dear Herr Professor!
I haven't burdened you for a long time now, and must now disturb you in your summer rest. I must ask you to read through both short English papers by Jones, because I don't understand much of them and Toronto is too far away to have Jones do the proofreading himself.[43] Please also read the French sections in Harnik once more.[44] I hope you are satisfied with the issue. There are a number of weaker brainchildren therein, but also some important contributions.

Issue I has already gone to Bergmann. I open with Nelken, then Wulff, Dattner and a number of interesting communications. You will get the proofs in time, so we can still plan changes. In Vienna we then want to take up the second issue.

I will remain here until the 12th and will travel to Wiesbaden through Vienna. I shall prepare Bergmann favourably for our project and try to convince him to come to Wiesbaden, so that the new affair may be consolidated. I have enough to do here. Unfortunately almost nothing psychoanalytical.

I hope you have recuperated well. They say I look very well and I feel very good.

With many cordial greetings
Yours loyally
Stekel

[20] Stekel to Freud
Zentralblatt für Psychoanalyse
Ischl, 29.8.1911

Dear Herr Professor!
I opened the first issue with Nelken, Wulff and Dattner. Elllis' paper is now in my hands. Should I not plan a change and replace Nelken with Ellis, Dattner with Ferenczi?

I prefer not to include the article. Many of our readers might take offence by it.

With regard to the changes, bear in mind that the first arrangement has already been printed, and that the publisher will incur considerable expense

[43] Presumably: Jones (1911a). Ein offener Brief. *Zentralblatt, I* (12), p. 610, and Jones (1911b). Darwin über das Vergessen. *Zentralblatt, I* (12), p. 614.

[44] Presumably the two short papers in the "Miscellaneous" [*Varia*] section of the *Zentrallblatt, I* (12), 612–613, on symbolism.

if it is changed. On 11 September shall leave for Vienna for a few days, then via Wiesbaden to Weimar.

With cordial greetings
Yours loyally
Stekel

[21] Stekel to Freud
ZENTRALBLATT FÜR PSYCHOANALYSE
Vienna, 15.9.1911

Dear Herr Professor!
I only received the *Jahrbuch* just now and am instantly and greatly impressed by your paranoia work.[45] Many of the new ideas were already present in me as seeds, as I pointed out in my little work on exhibitionism. Nevertheless, it is so new and overwhelming that I must speak with you about it.

The first issue is ready and I must send the second to the printers, because I want to make absolutely sure that we always appear on the first of every month. I will just leave open for the moment the societies section because I absolutely want to include Weimar and its resolutions in the second issue. So Rank will have to work very quickly. I propose for issue 2 as original contributions: Ellis, Luzenberger, Silberer; communications from Jones, Meissl, Stärcke and Stekel, also the old reviews, so that we finally get all the old materials done. In Weimar I then want to begin with a new layout of things. I am thinking of calling a meeting of the editorial board there.

Finally a personal note: I heard in Ischl that these days you are celebrating your silver [wedding] anniversary.[46] I wish with all my heart that you and your dearest ones will become living proof of my symbolic equation (silver–gold). Please send the entire family my cordial Congratulations.

When do you arrive in Weimar?[47] I travel to Wiesbaden tomorrow. My address will be forwarded to Bergmann the publisher.

Another note: It struck me how little the *Jahrbuch* elaborates on research from the Vienna school (with the exception of you!).[48] Coincidence or intention?

When will Ferenczi's work arrive? If it is to be published in the second issue, I should have it on my desk soon.

With many cordial greetings
Yours loyally
Stekel

[45] Freud (1911c). Psychoanalytic Notes on an Autobiographical Account of a Case of Paranoia (the Schreber case), first published in the *Jahrbuch*, III, part 1.

[46] Freud married Martha Bernays on 14.9.1886; their silver wedding anniversary took place on 14.9.1911 at Klobenstein.

[47] Weimar: third psychoanalytic congress, 21–22 September 1911.

[48] The *Jahrbuch*, III, Part 1, appeared in August 1911 and contained 15 contributions all of which were by Swiss analysts excepting two papers by Freud, one by Rank and another by Ferenczi.

[22] Stekel to Freud
ZENTRALBLATT FÜR PSYCHOANALYSE
Vienna, 5.10.1911

Dear Professor!
Please sign the addenda to our contract and return them to Bergmann.[49] As you can see, Jung's latitude has not been limited and the price of 12 Marks has been accepted. Now to hope for vigorous development. Subscriptions are at present very meagre. We must first publish the older stuff. Have you read Kovács' work on introjection?[50] I would like to include it in the next issue, with Sadger (migraine),[51] Reitler and Storfer.[52]

With cordial greetings
Yours affectionately
Stekel

[23] Stekel to Freud
ZENTRALBLATT FÜR PSYCHOANALYSE
Vienna, am 7.10.1911

Dear Master!
Please send the three contracts to Bergmann. He will stamp them.

I would be most grateful for any paper, however short. We can't afford to publish a second volume that is not up to the first.

Meanwhile a few short pieces arrived from Juliusburger and Hitschmann. Still, I expect massive assistance from Ferenczi, Jones and Maeder.

Please be cautious Wednesday. The Adlerians are planning to resign after all, as Rosenstein told me. It would be a tactical error if we were to allow them such a nice dramatic retreat. At any rate, we shall start with the congress report and the *Zentralblatt*.[53]

I recruited a new supporting member.[54] That is from Rank's point of view quite necessary.

[49] Presumably a renewal of the contract to publish the *Zentralblatt*.

[50] Kovács (1912). Introjektion, Projektion und Einfühlung. *Zentralblatt, II* (5), 253–263; (6), pp. 316–327.

[51] Sadger (1912). Über sexualsymbolische Verwertung des Kopfschmerzes. *Zentralblatt, II* (4), pp. 190–197.

[52] Presumably: Reitler (1912). Eine infantile Sexualtheorie und ihre Beziehung zur Selbstmordsymbolik. *Zentralblatt, II* (3), pp. 114–121 and Storfer (1912). Jungfrau und Dirne. *Zentralblatt, II* (4), pp. 200–204.

[53] At the meeting of the Vienna Psychoanalytic Society of 11 October 1911 Adler's followers announced their resignation after they had been confronted with Freud's injunction that one cannot be a member of both societies at the same time. Adler himself had already resigned earlier (*Minutes*, Vol. II, p. 146).

[54] Unidentified.

With many cordial greetings
Yours loyally
Stekel

[24] Stekel to Freud
MED. DR. WILHELM STEKEL
I. Bez. Gonzagagasse 21
Vienna, 15.12.1911

Dear Herr Professor!
I shall write to Assagioli according to your arrangements immediately. I have the feeling that he is not going to participate.

A nice paper from Abraham (the Weimar lecture) arrived.[55] It was about time, because we didn't really have any medical papers for issue no. VI (Issue 5 (Jones, Stekel, Kovacs) has already gone to the printers). Can I count on a short technical paper from you for issue VI? You have four weeks after all![56]

Friedjung also promised to write a practical article, for which I proposed: What does psychoanalysis have to offer children's medicine?[57]

In conclusion a request. A young lady from Trieste will report to you, she suffers from fear of fire and asthma. She is being treated by Hofbauer.[58] I would be most grateful to you if you would refer the young lady to me. She speaks with an Italian accent.

With cordial greetings
Yours loyally
Stekel

[25] Stekel to Freud
MED. DR. WILHELM STEKEL
I. Bez. Gonzagagasse 21
Vienna, 22.2.1912

My Dear Professor!
Herr Teirich has quit, as the enclosed letter shows.[58] Thank you very much for the good will.

[55] Abraham (1913). Ansätze zur psychoanalytischen Forschung und Behandlung des manische-depressiven Irreseins und verwandter Zustände. *Zentralblatt, II* (6), pp. 302–315.

[56] Freud contributed a paper to the 6th issue of the second volume: Über neurotische Erkrankungstypen. *Zentralblatt, II* (6), pp. 297–302 (see Freud, 1912c).

[57] Perhaps: Friedjung (1913). Ein Beispiel einer kindlichen Phobie. *Zentralblatt, II* (19–11), pp. 604–605.

[58] Unidentified.

I wrote to Riklin and Jung today. Schrötter has not yet shown up at my place.[59] He must be handled carefully. He shall certainly want to rank himself among our opponents.

With cordial greetings
Yours loyally
Stekel

[26] Stekel to Freud
ZENTRALBLATT FÜR PSYCHOANALYSE
Vienna, 29.12.1911

Dear Professor!
Your colossal Christmas gift in the shape of Dr. Handl arrived and had a visibly positive effect on the mood in the house. Please accept my hearty thanks. I forward to you a very pleasing enrolment. Sanitätsrat Gerster[60] is, as the enclosed work shows, a clever man and should become a strong co-worker. Löwenfeld befriended a Dutch psychotherapist through me. I referred him to Van Emden and Stärcke.[61]

I also mean to admonish you! On 10 January I shall send the March issue to the printer, which will comprise Freud, Abraham, Kovács (II.) (original contributions), Reitler, Robitsek and [Hug-]Hellmuth, and a new Swiss (communications).[62]

I hope you will be pleased.

With cordial greetings and many good wishes for the New Year
Yours loyally
Stekel
PS. I shall now have to call you Herr "Professor" in the *Zentralblatt*. Otherwise the truth will come to light that you are *the* professor, the others just *a* professor. It would perhaps produce sensitivities. (see Bleuler!)

[27] Freud to Stekel
VIENNA IX BERGGASSE 19
11.3.1912

[59] Presumably Karl Schrötter, who examined dream symbols experimentally and confirmed Freud's theory "more or less against his will" (see Freud to Jung, 18.2.1912). A paper of his on "experimental dreams" appeared in 1912 in the *Zentralblatt, II* (12), pp. 638–646.

[60] Unidentified. Gerster did not become a member of the board of the *Zentralblatt*.

[61] Jan van Emden (1868–1950) and August Stärcke (1889–1954) became president and member, respectively, of the Dutch Psychoanalytic Society in 1917. Van Emden was a frequent visitor of the Freuds and also briefly member of the Vienna Psychoanalytic Society. The identity of the psychotherapist is undisclosed.

[62] *Zentralblatt* I (6) contained the following papers: Freud, Ueber neurotische Erkankungstypen; Abraham, Ansaetze zur psychoanalytischen Erforschung und Behandlung des manisch-depressiven Irrsinns; Kovács, Introjektion, Projektion und Einfühlung (part 2).

Dear doctor
It is touching of you, and a disgrace for contemporary mankind, that you wrote almost all the reviews in No. 8.

I only have one comment: in the review of Moll, the last paragraph (How it ...) *must* be cut out. Moll is a snake, who deserves nothing more than a dry review and who will be unusually pleased with your demand that one of us must find a place in the second edition. He is not someone with whom one can curry favour.

Furthermore, I believe that the Miscellaneous section is too full, and that we should instead include more original contributions.[63]

Cordially
Yours
Freud

[28] Stekel to Freud
ZENTRALBLATT FÜR PSYCHOANALYSE
Vienna, 10.4.1912

Dear Professor!
Please find enclosed my contribution for the masturbation discussions, in which I struggle against various misconceptions. It does not express all I wanted to say, but it should counterbalance various voices we have heard in the discussions. But is certainly not—my last word on this issue.

I received Jung's photograph. If there is going to be no congress, there won't be a celebration issue either. What shall we do? Should we keep the photo until next year or do we publish it without a celebration issue? I believe the latter would be in bad taste. After a period of fairly exhaustive work, I now have a lot of time on my hands again. Don't you have a case of some kind? I would be most grateful.

I have now read through *Imago*. Pfister is impressive, but he doesn't contribute anything new.[64] Your paper is, like everything from you, daring, and it ought to generate much new work.[65] The rest is really nice, but I would have preferred a few popular tunes.

With cordial greetings
Yours loyally
Stekel

[63] Volume II, issue 8, of the *Zentralblatt* carried original contributions by Abraham, Rank, and Silberer; there were two communications and 9 reviews, 6 of which were written by Stekel. Stekel's review of A. Moll "Handbuch der Sexualwissenschaften" did not carry the said phrase.

[64] Pfister (1912). Anwendungen der Psychoanalyse in der Pädagogik und Seelsorge.

[65] First part of Freud (1912–1913). *Totem and Taboo*, first published in *Imago*.

[29] Stekel to Freud
ZENTRALBLATT FÜR PSYCHOANALYSE
Vienna, 12.5.1912

Dear Professor!
Through your good graces Frau Aronsohn is now under my care. It bodes well that her doctor in Riga already mentioned me and showed her my book on anxiety. The doctor in question appears to be very sympathetic towards psychoanalysis. I shall write him a publicity letter. One could start a Russian organisation now. I myself have about twenty followers in Russia and now the group in Moscow, etc.[66]

Did you read the works of Marcinowski and Juliusburger?[67] Which one should we include in issue 12? I want to open with Spielrein's paper, which is very marvelous and will cause quite a stir.[68] It is a document, and as such important. From Ferenczi a new piece of "miscellaneous" came in.[69] We now have an abundance of such small pieces and we have sufficient longer pieces until November. Shorter *articles* would be very welcome—particularly for the first issue. I mean of course papers from Freud, which are much in demand. Oh!—Recently a German baroness arrived and I am now fully booked. With cordial greetings and many thanks ...

your loyal pupil
Stekel

[30] Stekel to Freud
ZENTRALBLATT FÜR PSYCHOANALYSE
Vienna, 21.5.1912

Dear Herr Professor!
I am not so fond of secrets and I want to express myself openly. The perpetual attacks from Tausk make it simply impossible for me to participate actively in the Wednesday evenings meetings, which are no longer a pleasure. I can only take part in them the way I used to on condition that Dr. Tausk gives you the solemn promise that he will not mention my name in this Society at all any more and refrain from entering into discussion with me (not even friendly ones). I had wanted to press for this demand months ago, but then dropped it because I had hoped Herr Tausk would realize how impertinent and ridiculous his behaviour is. *Now I must insist on it.*

[66] Stekel's work was discussed by Ossipow in the Russian journal *Psikoterapiia* (Psychotherapy), of which Stekel became a member of the board (see Miller, 1998).

[67] Marcinowski (1913). Drei Romane in Zahlen. *Zentralblatt, II* (12), pp. 619–638; Juliusburger (1913). Beitrag zur Psychologie der sogenannten Dipsomanie. *Zentralblatt, II* (10–11), pp. 551–557.

[68] Spielrein (1912). Beiträge zur Kenntnis der kindlichen Seele. *Zentralblatt, III* (2), pp. 57–72.

[69] Presumably: Ferenczi (1913). Ein Fall von "déjà vu". *Zentralblatt, II* (12), pp. 648–649.

You may find my claim ridiculous and impudent, but I cannot do otherwise, and have given sufficient proof of my indulgence and placability. Should however there be another attack of this kind, I will regrettably be forced to draw the ultimate consequences. With cordial greetings

Your ever loyal
Stekel

[31] Stekel to Freud
DR. MED. WILHELM STEKEL
Vienna I, Gonzagagasse 21
1.6.1912

Dear Herr Professor!
I need to talk to you urgently. We will have to go through the collected material for the *Zentralblatt* and draw up a general plan for the third volume. It is hardly likely that we shall meet before October, and the first issue must already be ready by then. In addition, the masturbation debates contain much matter for discussion. I am now reading that terrible book by Rohleder on masturbation.[70] Only then does one notice the enormous progress our discussions will reveal. Finally, there are so many undecided psychoanalytic problems to discuss. When shall I see you to discuss this? I oftentimes wished I lived in Odessa or Hamburg, and could come to Vienna for a week. Nothing alienates more than proximity ...

I find it more and more difficult to speak in the Society. I do not feel surrounded by friends. It has actually become terribly uncomfortable for me, although I have certainly done everything I could to make friends. But the jealousies... . Enough of it!

So, when can we get together for an evening? Perhaps during the holiday?

With cordial greetings.
Yours loyally
W. Stekel

[32] Stekel to Freud
DR. MED. WILHELM STEKEL
Vienna I, Gonzagagasse 21
25.6.1912

[70] Rohleder (1899). *Die Masturbation*. Rohleder argued, according to Havelock Ellis in his chapter on autoeroticism in *Studies in the Psychology of Sex* (1927), that "masturbation may injure mental capacity, by weakening memory and depressing intellectual energy; that, further, in hereditarily neurotic subjects, it may produce slight psychoses like *folie du doute*, hypochondria, hysteria; that, finally, under no circumstances can it produce severe psychoses like paranoia or general paralysis."

Dear Herr Professor!
Unfortunately I could not meet with you. I leave tonight and must take my leave of you this way. I hope you are satisfied with issue XII. You will be able to see issue no. I of the next volume too. We shall edit the masturbation discussion in harmony. I shall keep you informed of current events. Please always forward your address to me.

I live in Ischl, Kaltbachstrasse 26, the same address as last spring.

I have three cases to psychoanalyse. I do not, however, wish to take on more because I am tired and yearn to go on holiday. I shall go to Lake Garda this September.

Yours loyally
Stekel
Please do not forget that I want to open the first issue with a small technical paper from your pen.

[33] Stekel to Freud
DR. MED. WILHELM STEKEL
Vienna I, Gonzagagasse 21
Ischl, 29.6.1912

Yesterday I had a consultation with Dr. Steinsberg and examined your mother. It is a bad case, but definitely progressing, which, however, the terrified patient would never admit. I have the impression that too much medicalisation has taken place. The patient is clearly against medication and very sensitive to anaphylactic reactions [allergic response]. On the other hand, many harmless physical medications were left untried. I recommended constant application of heat through large termophores [moist heat packs], radiation with red sunlight (sometimes soothes very quickly!) and other trifles, from which I imagine the best results.

I am constantly at your dear mother's disposal and hope to inform you of the good results of the new therapy soon.

With cordial greetings
Yours loyally
Stekel

[34] Stekel to Freud
ZENTRALBLATT FÜR PSYCHOANALYSE
Ischl, 8.7.1912

Dear Herr Professor!
I bring you favourable news about your dear mother. There is noticeable progress. She looks much better and praises my efforts.

I must also admonish you also. I need your contribution to the masturbation discussions. I have conscientiously read through all the other contributions and would prefer not to change a word. Repetitions are few, and when they are, they conform to Goethe's maxim: Full three times must it be ... .[71] So I beg you, please write your contribution right away.

I shall write the preface these days and send it to you.[72]

Sitting behind my typewriter I can see Reitler on mama's balcony ... .

I am doing very well. I do not do anything literary or analytical at all only three patients.

With cordial greetings
Your devoted pupil
Stekel

[35] Stekel to Freud

ZENTRALBLATT FÜR PSYCHOANALYSE
Ischl, 14.7.1912

Dear Herr Professor!
I will open the new volume with Oppenheim, [Hug-]Hellmuth and Spielrein.[73] Do you agree? Then I plan to write an interesting communication: the representation of neurosis in dreams. By that I mean those symbols which represent the neurosis personified. (Virgin soil)

With respect to our priority struggle: I argued against the claim, put forward by Ferenczi, that it holds for all cases, and gave a different interpretation. I have the essays in Vienna ... Incidentally, you are right. I read the passage again. There it says: *This* explanation was until now etc... .[74]

We will have to forgo *Steiner*. I cannot hold back the booklet only because Steiner cannot or does not want to write. With you, it is an entirely different case. You have, after all, the concluding speech.

Dear mother was doing very well. Yesterday she had a bad night, which can probably be traced back to too much galvanization. I recommended innocent measures. Variatio delecate. Your sister reveals nice conversion

[71] Goethe, *Faust* (Part I, "In the study"). Faust has to invite Mephistopheles into his study three times to symbolize his willingness to become involved in the evil the spirit represents. Faust: "A knock? Come in! Who now will bother me?" Mephistopheles: "Tis I". Faust: "Come in"! Mephistopheles: "Full three times must it be."

[72] A preface ("Zur Einleitung") to the masturbation discussions was published under the collective authorship of "the editors of the Vienna psychoanalytic discussions."

[73] Vol. II, nr. 1 of the *Zentralblatt* opened with Juliusburger and Hug-Hellmuth.

[74] Only part of this discussion can be reconstructed as pertaining to the masturbation debate. Ferenczi claimed in his contribution that a man who has intercourse with his wife but phantasizes about another woman during the act masturbates "in vaginam." Stekel contradicted this finding in his contribution and excluded a priori all behaviour that involved another party as masturbatory: "masturbation is by definition an asocial act." It is unclear, however, how this relates to a "priority struggle" (with Freud), nor were we able to trace the sentence about how 'this explanation' relates back to this matter.

symptoms. She was going to tell Dr. Steinsberg that he need not return, as Reitler is taking over treatment. But she did not have the courage to do it. Instead, she got a terrible sore throat and Dr. Steinsberg had to come to examine her throat and found ... nothing. The words got stuck in her throat and she wanted Dr. Steinsberg to get them out (to guess it) ... She cleared her throat and asked: You really see nothing? In the evening she came to me: I can't find the words... How shall I say it to the doctor? Why doesn't he understand himself?

I am truly pleased with my new book *The Dreams of Poets*.[75] But what will my Viennese colleagues say? I would like to prevent a discussion about it in the Society. I shall speak with you about it in private, but I do not care much about what the rest have to say about it.

With many warm greetings, best wishes for a pleasant holiday which you should not interrupt to send me a reply.

Yours loyally
Stekel

[36] Stekel to Freud

ZENTRALBLATT FÜR PSYCHOANALYSE
Ischl, 18.7.1912

Dear professor!
The enclosed contribution is from Hitschmann.[76] He would like your opinion and judgement. I find it somewhat weak but interesting and would like to print it.

What is your view?

Otherwise no news. Dear mama, who is now being treated by Reitler, is improving.

With cordial greetings
Yours loyally
Dr. Stekel

[37] Stekel to Freud

ZENTRALBLATT FÜR PSYCHOANALYSE
Ischl, 29.7.1912

Dear Master!
Time passes and Steiner's essay has not yet arrived.[77] We cannot wait any longer. Unfortunately, as editor I have to press you for your concluding

[75] Stekel (1912). *Die Träume der Dichter*. Wiesbaden: Bergmann.

[76] Perhaps: Hitschmann (1913). Swedenborg's Paranoia. *Zentralblatt, III* (1), pp. 32–36.

[77] Steiner's contribution to the masturbation discussions appears as the next to last chapter in the volume of the "Vienna discussions" on masturbation.

word and urge you to write it as quickly as possible. I insist that we appear in the autumn as planned so that the fine discussions showcase Vienna as the leading force.

Issue no. I of the third volume has gone to the printers. You will be satisfied. I hope you will be pleased with my contribution on "The personification of neurosis in dreams". A new symbolism! Additional contributions from Hitschmann, Ferenczi, Jones and many others. Very extensive, nice reviews. Unfortunately, Reitler has not sent me the *Jahrbuch* reviews. People are lazy! Where would I be today if I were so easy-going!

What is going on in Zurich? Please send me instructions. And where may my letters reach you?

Mama is actually doing very well.

With many cordial greetings
Yours loyally
Stekel

[38] Stekel to Freud

ZENTRALBLATT FÜR PSYCHOANALYSE
Ischl, 31.7.1912

Dear Herr Professor!
Thank you very much for your expression of sympathy, I can send you a reassuring message. The dog was only under suspicion and the vaccination was only a precautionary measure. But you can imagine the excitement... .

I heartily congratulate you on the engagement of your daughter.[78] I do not need to send you the customary good wishes. You know how much you mean to me and the part I play in your history. In the history of psychoanalysis our names are closely linked and I remain the prophet, you the Allah!

Please do not forget the concluding article.

Ceterum... [79]

I am proofreading my new dreambook[80] and already look forward to sending it to you soon.

With many cordial greetings
Yours loyally
Stekel

[78] Engagement of Sophie Freud.

[79] *Ceterum censeo Carthaginem esse delendam*: "Therefore I conclude that Carthage must be destroyed" (Marcus Porcius Cato).

[80] Stekel (1912). *Die Träume der Dichter*. Wiesbaden: Bergmann.

[39] Stekel to Freud
MED. DR. WILHELM STEKEL
I Gonzagagasse 21
Vienna [no date, Summer 1912]

Dear Professor!
Thank you for sending me [the 4th edition of] *The Psychopathology of Everyday Life*.[81] The book is getting richer and better every time. I want to review it in the next issue. It is made for propaganda for our cause.[82]

In the rich garland of blossoming flowers with which you decorated your friend's Ferenczi's head I found a little violet I claim to be mine. Not in public! God forbid. Just to call to your own mind, which seems to show a tendency to forget me. The *first* who searched the topic "déjà vu", and who spoke and wrote on it in public bears the name Stekel. Did you forget that I gave a speech on this subject in your apartment (Kahane era), and that we discussed it for two consecutive evenings? I then wrote in *Nord und Sud* a lengthy paper entitled "I have experienced that before", also in *Tagblatt*, etc... .[83]

But all that is not important. I am sending you the introduction, or rather the design of the introduction to the [masturbation] "Discussions".

Please change whatever you do not approve of and add whatever you find necessary.

Mother is making a much better impression, the distension in her hand has rapidly decreased. She still complains of pain, I believe more as a "point d'honneur", so as not to make the cure look "bad".

I wish you a complete recovery in Karlsbad. I am very well.

With many cordial greetings
Yours loyally
Stekel

[40] Stekel to Freud
ZENTRALBLATT FÜR PSYCHOANALYSE
Ischl, 17.8.1912

Dear Herr Professor!
I just read your contribution to the masturbation discussions and found it quite excellent. I feel honoured that you busy yourself with me so much. You levelled a small injustice at me, however, at the end. Or a significant lapse? You believe forms of unconscious masturbation were not sufficiently examined, but I was the only one who dealt with this fairly exhaustively ...

[81] Freud (1901). The Psychopathology of Everyday Life, *S.E.* 6.
[82] Stekel (1912). Review of Freud, Psychopathologie des Alltagslebens. *Zentralblatt, III* (1), p. 47.
[83] Unidentified.

I just read Furtmüller's *Psychoanalysis and ethics*.[84] It is written quite decently. But what do you think of the introduction? Adler's book is terrible ... [85] I don't know what to make of it.

Have you received the proofs of the first issue yet? I am very curious to see how the subscription rate will fare and the extent to which *Imago* is going to harm us. Issue III is already in my hands and appears to be very copious. The works on colour tone belong to the *Zentralblatt*.

When do you return to Vienna? I have many urgent matters to discuss with you. Vienna must come first, then the problem of the congress, the Viennese relations, the position towards the counter emperor (Adler), which must no longer be a passive one, etc. Furthermore, we must deal with internal relationships in the Society, distribution of work, styling of the *Zentralblatt*. I've heard nothing at all from the *Korrespondenzblatt*. I [illegible] the same with Riklin. With many cordial greetings,

yours affectionately
[Stekel]
P.S. Yesterday, I found Frau mama in good condition. Reitler makes a strange impression on me ... good-humoured ... but lame.

[41] Stekel to Freud
ZENTRALBLATT FÜR PSYCHOANALYSE
Ischl, am 28.8.1912

Dear Herr Professor!
Has a "good friend" already send you that foul feuilleton by Reich in the *Frankfurter* [*Zeitung*] on Adler?[86] If not, I will gladly send it. I hope you can laugh about the mix of stupidity, viciousness and intellect. I suddenly feel like going to the congress in Zurich. Seif, Jones, Maeder and Adler will be there. There will be discussions on anxiety neurosis. What do you think?

Dear mama is much better. I warmly recommended Aetherspray, it seems to work very well to relieve the pain. I recommended, furthermore, inhalation of a few drops of chloroform, a known remedy to resolve congestion.

Not much other news, at most a daily shower. I support the idea that mama might go to Meran or Abbazia ...

With many cordial greetings
Yours loyally
Stekel

[84] Furtmüller (1912). *Psychoanalyse und Ethik*. Schriften des Vereins für freie psychoanalytische Forschung. München: Reinhardt.
[85] Adler (1912). *Über den nervösen Charakter. Grundzüge einer vergleichenden Individualpsychologie und Psychotherapie*. Wiesbaden: Bergmann.
[86] Unidentified.

[42] Stekel to Freud
Zentralblatt für Psychoanalyse
Ischl, [?] Aug. 1912

Dear Herr Professor!
You got me wrong! I don't want to pick a fight with Adler. In my latest book you will hunt in vain for any such controversies. The best fight is in the proof that things are not how Adler sees them ...

Just a while ago your sister was here and complained that mama has had some very bad nights. The poor woman suffers from a lack of Tarok games, a lack of entertainment and a lack of sunshine. We have terrible weather over here, too much rain, and mama seems unable to tolerate it well. I will meet with Reitler tomorrow and discuss a few options. But I must be very careful here so as to spare the sensitivity of R.'s colleagues and friends. I do believe, however, that a sojourn in Abazzia or at the Garda lake would do mother much more good than a stay at Ischl. Perhaps it would be best if she went to Vienna's health resort, where she would have more diversions and less humidity, which doesn't do her any good.

I have reassured your sister and hope that I shall be able to send you a favourable report soon.

With many cordial greetings
yours loyally
Stekel
Silberer was with me today. And Hitschmann too was in Ischl. Soon we'll be able to start a branch Society in Ischl ... [87]

[43] Stekel to Freud
Zentralblatt für Psychoanalyse
Ischl, 11.9.1912

Dear Herr Professor!
Since the weather is persistently bad, I advised dear Mama to travel to Riva. I found wonderful accommodation with Dr. Hartungen.[88] She is now only waiting for permission from your brother ... [89]

I took a look at the third issue of *Imago* and regretfully saw that it includes purely medical works which belong to the *Zentralblatt*. I mean papers by Hug–Hellmuth and Pfister on colour tone. One of those two could have been published in the *Zentralblatt*. That would have been more sensible and more agreeable for readers of both. I would find such an editorial policy of *Imago* irritating and it makes me wonder. I would like to ask

[87] In his autobiography Stekel (1950a, p. 126) claimed that he lived five months of the year in Ischl and practiced psychoanalysis as regularly there as he did in Vienna.
[88] Unidentified.
[89] Presumably Alexander Freud (1866–1943).

you to use your influence so that such intrusions in our territory be prevented. *Imago* only has the right to exist if it passes to the *Zentrablatt* all medically or psychologically oriented papers. On the other hand, we shall pass all literary and artistic works to *Imago*... If not, I shall have to conflict openly with *Imago* and include an excellent literary paper in the next issue. You'll understand my viewpoint if you put yourself in my position ... I leave Ischl on the 15th and travel back to Vienna.

Hard work awaits us.

With many cordial greetings
Yours loyally
Stekel

[44] Stekel to Freud
ZENTRALBLATT FÜR PSYCHOANALYSE
Vienna, 14.10.1912

Dear Professor!
Thank you very much for sending me the interesting paper, and thank you even more for your promise to write a short technical article every other month. After we've seen the *Zentralblatt* through its third year, it should be able to flow on its own on the high tide.

But what to do with the review commission. Since Tausk has signed himself up for it, my cooperation is out of the question. I already can hardly bear that this man is in our Society. The Zuricher too should only be judged by their *best* workers. You are entitled to that! A few things I wish to consider myself, such as Nelken's work, or Silberer's work. But how to do this when this [character] Tausk is present, with whom I cannot even sit at the same table under any condition. Maybe I shall suggest the colleague give up his chair in the committee in my favour, or forgo it altogether, for the reasons just mentioned ...

I find it unpleasant to tire you with these internal troubles, but I have been pushed into this position and reconciliation [with Tausk] now or later is out of the question. We need all the forces we have, now more than ever ... I do not want to pick a fight, only some consideration of the present situation. Frankly: every review from Tausk in my journal would spoil for me the pleasure of working on my beloved *Zentralblatt*.

I know you can solve this situation with just a few words, internally, without debate.

Another chapter is the problem of Tausk's learning capacity. I believe he has still a lot to learn, not in the least with regard to dissolving his own complexes.

With many cordial greetings
Yours loyally
Stekel

[45] Stekel to Rat
MED. DR. WILHELM STEKEL
I Gonzagagasse 21
Vienna, 17.10.1912

Dear Herr Rat!
Many thanks for your kind words. Payment is out of the question. The allusion alone is an affront. Surely I can do the mother of my teacher a small favour?

I understand that your dear mother is much better. I am happy about that.

With many cordial greetings
Yours truly
Stekel

[46] Stekel to Rat
MED. DR. WILHELM STEKEL
I Gonzagagasse 21
Vienna, 21.10.1921[90]

Dear Herr Rat
You find me in an awkward situation. I have never thought about [asking] a fee and never booked the number of visits I made to Frau Mama [Freud].

Next time my assistance is called upon I shall assuredly do so, and I herewith consider this business concluded.

In confidence! I can probably settle matters with the brother. He seems unable to recognize and *keep* his true friends ...

Please do not talk about this. More by word of mouth some time later.

With many cordial greetings
Yours truly
Stekel

[47] Stekel to Rat
MED. DR. WILHELM STEKEL
I Gonzagagasse 21
Vienna, 23.10.1912

Dear Herr Rat![91]
You seem to have mistaken me for Dr. Reitler, who was responsible for daily supervision. I cannot accept such a large fee for those few consultations.

[90] Handwritten in pencil across the letter twice: SCHWEIN ("Swine"). In the margins two exclamation marks appear and the words "In confidence" and "Please do not talk about this" are underlined At the bottom is a note in pencil on the expenditure: K[ronen] 300,-21/10 1912 transported [word unclear] Rezepine. Postal money order.

[91] An arrow in pencil points from the name "Rat" to the handwritten name "Alexander".

I will keep, so as to reassure your conscience, 50 kronen and return the remaining 250 with thanks.

Our business is herewith concluded.

With cordial greetings
Yours truly
Stekel
[A handwritten note at the bottom of the page:] What to do? Did not yet reply. I am in favour of accepting his proposal or else to split the sum of 300 K [ronen] in two, that is, 150 K [ronen].

[48] Stekel to Freud
[No date, ca. November 1912]

Dear professor!
Naturally your work will only be published when it suits you. The arrangement of the third issue is only provisional.

Most respectfully
W. Stekel

[49] Stekel to Freud
[No date, ca. November 1912]

Dear professor!
I inform you that for technical reasons the changes you requested in the *Zentralblatt* can only be processed as of the December issue.[92] The November issue was already printed weeks ago.

Respectfully
Dr. Stekel.

[50] Stekel to Jung
ZENTRALBLATT FÜR PSYCHOANALYSE
Vienna, 19.12.1912

Dear colleague!
As you can see from the enclosed envelope, you seem to have confused me with Sadger, whose address is on the cover. I do not consider that a compliment ...

You suggest in your letter that I and Adler forced a fracture. But you should know that Freud forced the break with Adler and even drove him and eight of his followers, who wished to stay with us, out of the Society because he confronted them with the choice: Freud or Adler.

Also, my conflict was not caused by Tausk. Freud had it in his power to settle the entire matter with merely a few friendly words. I would also

[92] Presumably the removal of the names of the collaborators to the *Zentralblatt*. The third issue of the third volume was headed with the following announcement: "Herr Professor Freud renounced as publisher of the journal."

have resigned from the management of the *Zentralblatt* if Freud had tried to bring the affair to an end in a friendly way. I even came to him after the break and offered him a peace treaty. But he was apparently put out by my latest book *The Dreams of Poets*. He cannot bear dissention from his teachings, and if he does, as with you, it is under pressure. But your hour too will come soon, it already has. My sacrifice restored peace between Zurich and Vienna for a short while. But for how long?

Did you not notice that Freud cannot keep friends? Are only those friends to blame in that. I went to the limits for my teacher in what is possible. I also shall never want to forget what I owe him. But I also know my commitment to the Freudian cause, which has not been small ...

I realize that you and many of your followers do not think much of my researches. Time will teach you to think better of them. Do read the latest critique in the November issue of the *Journal of nervous and mental disease*, and you must admit that in America they understand me better than they do here in my own small circle.

For me the matter is finished. I just wanted to let you know that a sharp critique of your *Transformations of the Libido* will appear in my journal. I gladly open the columns of my journal for a rebuttal. It is from such discussions that readers will learn: their own opinion. That which arises from dispute is thus a blessing for our movement. We can learn from one another and preserve ourselves from partiality.

I discovered so many new facts within the psychoanalytic domain that I need years to work them out.

The second volume of my "Disorders of the Instincts and the Emotions" series[93] will prove whether or not I am a hazard to psychoanalysis, as the favourite phrase puts it, which associates me with Sadger and Abraham ... .[94]

With many cordial greetings
Yours truly
Dr. Wilhelm Stekel

[51] Stekel to Freud
MED. DR. WILHELM STEKEL
I Gonzagagasse 21
Vienna 18.2.1913

Dear Professor!
I am proofreading the second printing of *Nervous Anxiety States* [*and their Treatment*]. As our relationship has changed in nature for some time, I allow the question of whether or not you wish to retain the introduction

[93] Stekel, *Störungen des Trieb-und Affektlebens: die parapathischen Erkrankungen*. Vol. II: *Onanie und Homosexualität* (*Die homosexuelle Parapathie*). Vienna: Urban & Schwarzenberg.

[94] Apparently this *was* a common equation in psychoanalytic circles; see letter of 14.1.1912 of Freud to Abraham: "Do you know that you count with Stekel and Sadger among the bete noires of psychoanalysis against whom I have always been warned."

which you so kindly wrote. Please inform me of your decision in an unreserved manner, as you have always done before.[95]

With cordial greetings
Yours truly
Dr. W. Stekel.

[52] Stekel to Freud
EROS UND PSYCHE
Vienna, 11.10.1915

Dear Professor!
I have been told that my conduct in [Bad] Gastein has upset you.[96] It is true; I did not greet you. I was afraid you would not return my salutation. Surely you have forgotten that shortly after we separated (on that night in café Landtmann) you never returned my salutations.

I would have otherwise considered it a natural duty to bestow my teacher and master the honour I owe him.

Be assured, however, that since I overcame my bipolar disposition, which led to defence partly because of your accusation of my having "gone astray"[97], I have no other feelings for you than honour and gratitude for the genius who lit the torch which illuminates the road of my endeavours.[98]

Yours truly
Dr. Wilhelm Stekel.

[53] Freud to Stekel
PROF. DR. FREUD
Vienna IX, Berggasse 19
13.1.1924

Dear Doctor
I acknowledge receipt of your letter of 31 December 1923[99] and thank you for your good wishes regarding the improvement of my health.[100] I cannot refrain from contradicting you on a few important points however.

[95] See Freud, 1908d. From the second edition onwards, Freud's preface was omitted from "Nervous Anxiety States", and replaced with a preface of Stekel in which part of Freud's piece was reproduced. Stekel suggested that he withdrew it himself, but this letter makes it possible that Freud refused republication.

[96] Mid July 1915 Freud went to Bad Gastein to spend his holidays there, but conditions were so unsatisfactory that he went over to Salzburg, only to return to Bad Gastein at the end of August (see Jones, 1953–1957, Vol. II, p. 189). Freud would run again into Stekel when he was in Bad Gastein in 1921.

[97] Reference to 'On the History of the Psychoanalytic Movement' (Freud, 1914, p. 19).

[98] The exact same phrase appears in the concluding remarks in the book on *Compulsion and Doubt* (Stekel, 1962a, p. 630)

[99] Letter missing.

[100] Freud was operated on 28 October and again on 12 November 1923 and underwent a Steinach operation on 17 November. He resumed work on 2 January 1924 (see Jones, 1953–1957, Vol. III).

You are mistaken if you think that I hate you or have ever hated you. The fact is that after an initial sympathy—perhaps you remember how our relationship began—I had reason for many years to be annoyed with you while at the same time I had to defend you against the aversion of everyone around me, and that I broke with you after you had deceived me on a certain occasion in the most heinous manner. (You never mention this occasion—*Zentralblatt*—in your letters.) I lost confidence in you at that time and since then you have not provided me with any experience that could help me regain it.

I also contradict your often repeated assertion that you were rejected by me on account of scientific differences. This sounds quite good in public, but it doesn't correspond to the truth. It was exclusively your personal qualities—usually described as character and behaviour—which made collaboration with you impossible for my friends and myself. Since you most certainly will not change—you don't need to, for nature has endowed you with an unusual degree of self-complacency—our relationship stands no chance of becoming any different from what it has been during the past twelve years. It will not annoy me to learn that your medical and literary activities have earned you success; I admit that you have remained loyal to psychoanalysis and have been of use to it; you have also done it great harm.

My friends and pupils will find it easier to value your publications objectively when you begin to voice your criticism and polemics in a more polite tone.

With best wishes
Yours cordially
Freud

[54] Stekel to Freud
DR. WILHELM STEKEL
Vienna 22.1.1924

Dear Professor!
You know from your own experience that there are always two sides to justice and that the same incident in the minds of two people can look very different. I had the intention of writing a history of the analytical movement. I did not execute this plan and waited until I had gained more distance from the events and I subsequently threw myself into scientific work.

I do not wish to speak of the past. You see only the injustice done to you and you overlook the mistakes you made. Had you acknowledged the sources from which the rivalry between your students arose in time, you could have retained many useful hands. It was not only a struggle of pretenders to the throne, but a competition for your love. It had more to do with jealousy for your heart than a claim on your head. I recognize similar tendencies in my own circle and we tried through analytical frankness to clear the conflict-laden air. So far with success.

Please—don't misunderstand me. I did not write to you for personal gain. I am entirely satisfied with my accomplishments. You speak of my self-complacency. For sure. I am one of the few happy analysts who is also master in the art of living, who cares little for what the small or big world has to say. And believe me: I have a strong sense of self-criticism and see all my faults sharper than any of my opponents, but I have the courage to look them in the face and attempt to amend for them.

I wrote to you because I want to give our relationship, to which I owe so much, a reconciliatory resonance. If you think that I treated your friends and pupils badly I should like to rigorously check my critique in the future and to submit them to impartial justice without asking anything in return. This is not going to be easy since many of your students digress into speculations and phantasms, while I remain on the safe ground of observation and experience.

Psychoanalysis lacks a counter voice in its own camp. There is no opportunity of going one's own way. There are already some 60 members in my organization of independent medical analysts because individual freedom of research is not hampered. We do not displace scientific conflicts onto personal affairs. We do not excuse as differences between character what is really a difference of opinion.

Yet we vouch for analysis, and we all stand on Freudian ground. Consequently, parallel work [between both groups] and alliances against the opponents and diluters of analysis is possible.

Perhaps, in conclusion of our personal relations and with regard to a better *scientific* relationship, you might want to give me the opportunity to see you on a Sunday or some other evening?

First condition: the past is not discussed. But that first condition is also the only condition ... [101]

Respectfully
Yours truly
Dr. Wilhelm Stekel
P.S. In as much as critiques are already in print, nothing can be changed. The *Jahrbuch* will not publish any critiques. We want to criticize through our work. How successful we are, you can soon find out for yourself.

[55] Stekel to Freud
VEREINIGUNG UNABHÄNGIGE ÄRZTLICHER ANALYTIKER IN WIEN
Vienna, 8.5.1931

Dear master Freud!
On the day the entire world honours the name Freud as a special celebration in the Book of History, your oldest pupil and admirer cannot fall short. You

[101] Ernest Jones (1953–1957, Vol. III, p. 102) discusses this letter; he is certain that the proposed meeting did not take place.

may scorn me for being ungrateful, and your resentments may not have disappeared, but as an analyst you know that every individuality has its own laws which it must obey. Today I stand before you representing of a society of analytical medical men comprising of some 130 members, scattered across the world.

We are all aware of what we owe you. We all wish you will be with us and mankind for a long time. And we wish to express on this day that there is nothing that separates us except technique ... and some other things, which in the light of the greatness of your accomplishments need not be mentioned at all.

Wistfully I think of the day we planted the seeds, when we jointly worked in preparation for the future. I can claim with justifiable pride that I, always as a psychoanalyst and based on your work, helped show the physicians the importance of analysis.

And so I beg you to accept my congratulations, and those of my followers, on your 75th birthday. The future will prove that we are all Freudians in the best sense of the word.

With utmost respect, yours truly
Dr. Wilhelm Stekel

[56] Stekel to Freud
PEMBRIDGE COURT HOTEL
London, 7.9.1938

Dear Professor
A gloomy destiny drove us both away from our working homes and brought us to London, where psychoanalysis still finds the appreciation it is denied in Germany.

I too shall spend the twilight of my life here. My books are on the index too, although fortunately they are available in translation.

I am convinced, however, that it is a transient epidemic and that all these petty houndings will not do away with your fame and virtue.

I have always been a champion of psychoanalysis and I have brought over many adherents even when seemingly I have been estranged from your school.

Yet I have always been proud to have been your first student and co-worker and my last two books *Interpretation of Dreams* and *Technique of Analytical Psychotherapy* result in an expression of gratitude to my master and leader.[102] These sentences are meant to welcome you to London and to wish you many many more years of undisturbed work.

Yours loyally
Dr. Wilhelm Stekel.

[102] The last sentence of *The Interpretation of Dreams* (Stekel, 1962b, p. 572) reads: "The seed which Freud scattered has sprouted richly, and we all have ample reason for gratitude". See also Stekel, "On the History of the analytical Movement" (chapter 8) for a commentary on this letter.

Appendix

Zentralblatt für Psychoanalyse. Medizinische Monatschrift für Seelenkunde

Issue	Vol. I	Vol. II	Vol. III	Vol. IV	Month
1	1910	1911	1912	1913	Oct
2			Freud resigns		Nov
3			directorship		Dec
4	1911	1912	1913	1914	Jan
5			*Zentralblatt*		Feb
6			published under		March
7			sole responsibility		Apr
8			of Stekel		May
9					June
10					July
11					Aug
12	Adler resigns editorship			*Zentralblatt* ceases publication	Sept

PEMBRIDGE COURT HOTEL

TELEPHONES
BAYswater 5177 Guests
„ 5178 Office

34, PEMBRIDGE GARDENS,
HYDE PARK, W.2.

London 7/IX 1938

Sehr geehrter Herr Professor:

Ein trauriges Schicksal hat uns Beide von der bisherigen Arbeitsstätte entrissen und nach London gebracht, wo die P. A. noch die Anerkennung geniesst, die ihr in Deutschland entzogen wurde.

Auch ich werde hier meinen Lebensabend verbringen, auch meine Bücher stehen auf dem Index, sind aber glücklicherweise noch in Uebersetzungen zugänglich.

Ich bin überzeugt, dass es sich um eine vorübergehende Epidemie handelt und dass alle diese kleinlichen Verfolgungen Ihren Ruhm und Ihr Verdienst nicht verdunkeln können.

Ich war immer ein Vorkämpfer der P. A. und

A.A. RECOMMENDED R.A.C.

habe ihr viele Amhänger geworben, wenn ich mich auch scheinbar von Ihrer Schule entfernt habe.

Aber ich war immer stolz Ihr erster Schüler und Mitarbeiter gewesen zu sein und meine beiden letzten Bücher"Fortschritte der Traum-deutung " und die Technik der P. A. " münden in den Dank an meinen Meister und Wegführer.

Diese Zeilen sollen Sie in London begrüssen und Ihnen noch viele viele Jahre ungetrübter Arbeitskraft wünschen

Ihr getreuer

Dr. Wilhelm Stekel

The last letter of Stekel to Freud, dd. 7 September 1938, written to welcome Freud to London, two months after he arrived there.

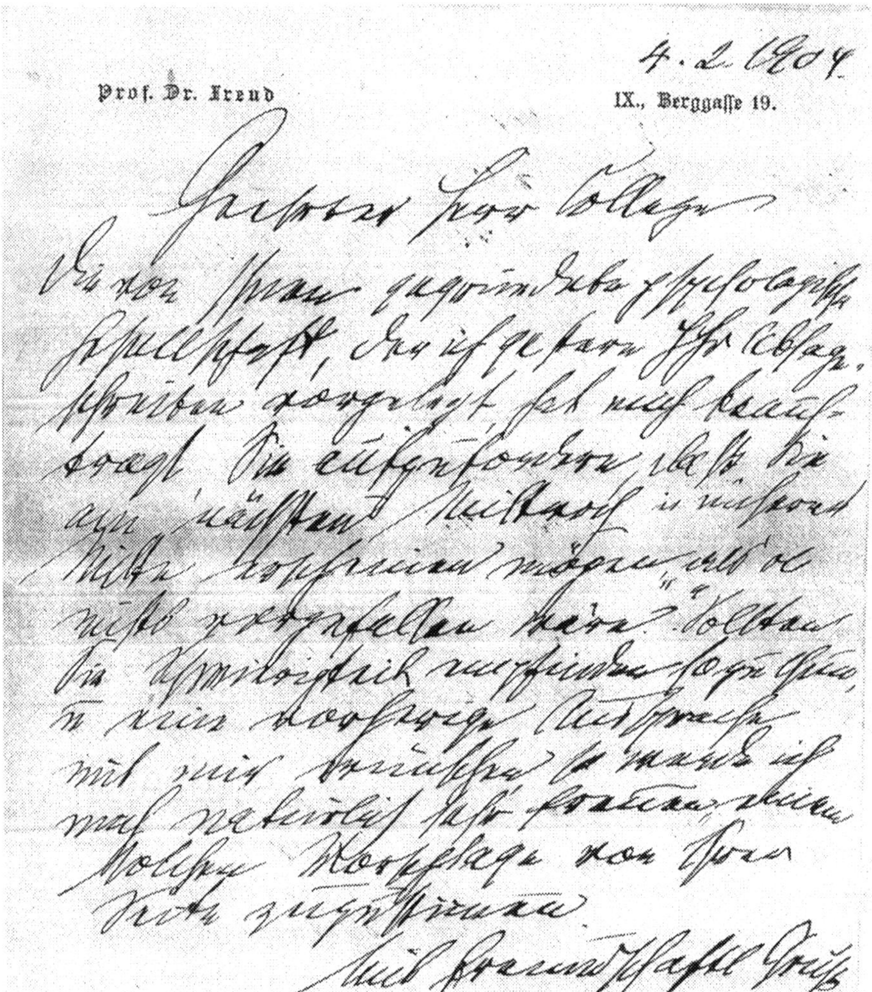
4. 2. 1904

Prof. Dr. Freud IX., Berggasse 19.

A 1904 note from Freud to Stekel, the first among the few letters of Freud to Stekel to have survived and the first evidence of their troublesome relationship.

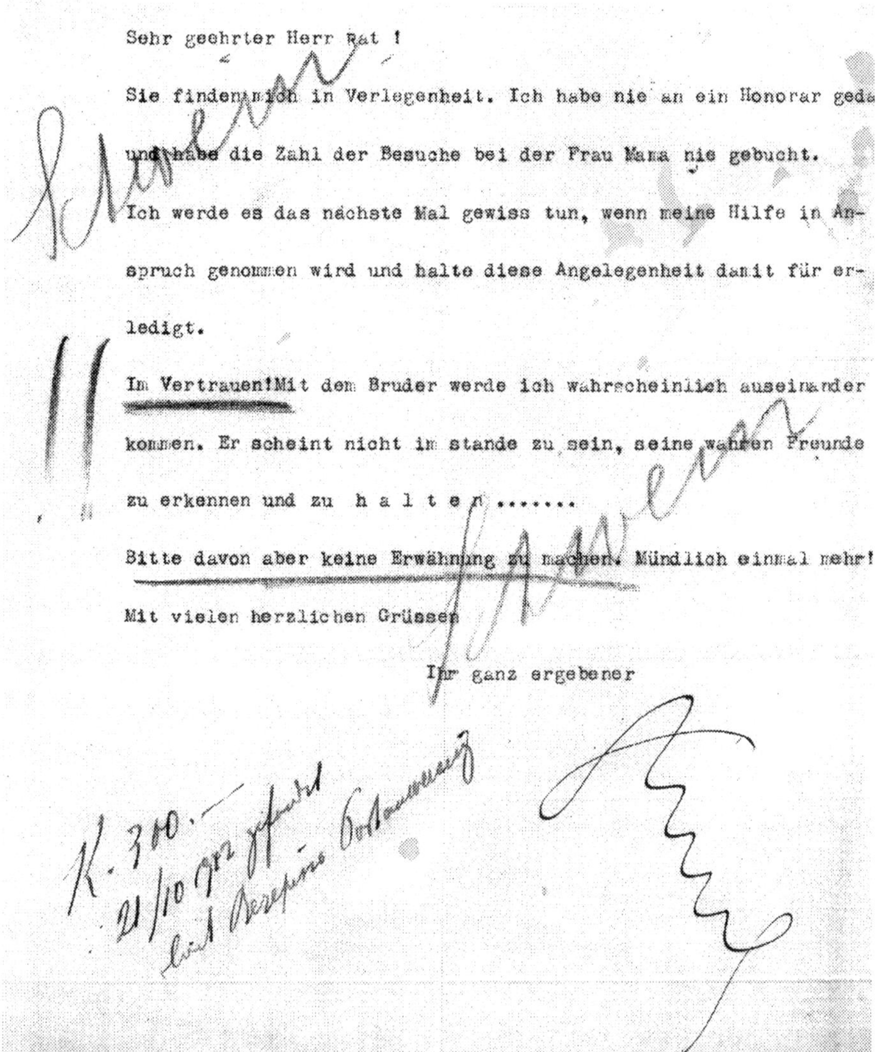

Sehr geehrter Herr Rat !

Sie finden mich in Verlegenheit. Ich habe nie an ein Honorar geda
und habe die Zahl der Besuche bei der Frau Mama nie gebucht.
Ich werde es das nächste Mal gewiss tun, wenn meine Hilfe in An-
spruch genommen wird und halte diese Angelegenheit damit für er-
ledigt.

Im Vertrauen!Mit dem Bruder werde ich wahrscheinlich auseinander
kommen. Er scheint nicht im stande zu sein, seine wahren Freunde
zu erkennen und zu h a l t e n ....

Bitte davon aber keine Erwähnung zu machen. Mündlich einmal mehr!

Mit vielen herzlichen Grüssen

Ihr ganz ergebener

The mysterious 21.10.1912 "Swine letter," a note of Stekel to a certain Herr Rat on the payment of Stekel's medical services lent to Freud's mother. The note precedes the break between Stekel and Freud by a week. At this point Freud was determined to get rid of Stekel and sought to pay off any dept he might have to separate from his colleagues as clean as possible.

References

Ackerman, N. W. (1958). *The Psychodynamics of Family Life. Diagnosis and Treatment of Family Relationships*. New York: Basic Books.
Adler, A. (1912). *Über den nervösen Charakter*. Vienna: Bergmann.
Adler, A. et al. (1910). *Über den Selbstmord insbesondere den Schüler-Selbstmord* (Diskussionen des Wiener psychoanalytischen Vereins, Vol. 1). Wiesbaden: Bergmann.
Albach, F. (1993). *Freud's verleidingstheorie. Incest, trauma, hysterie*. Middelburg: Stichting Petra.
Allen, G. M. (1959). The Adlerian Interpretation of Compulsion. In K. A. Adler & D. Deutsch (Eds.), *Essays in Individual Psychology. Contemporary Application of Alfred Adler's Theories* (pp. 55–58). New York: Grove Press.
Andreas-Salomé, L. (1987). *The Freud Journal*. London and New York: Quartet Encounters.
Bakhtin, M. (1981). *The Dialogic Imagination*. (M. Holquist, Ed.). Austin: University of Texas Press.
Barton Evans III, F. (1996). *Harry Stack Sullivan. Interpersonal Theory and Psychotherapy*. London/New York: Routledge.
Beech, H. R. (Ed.). (1974). *Obsessional States*. London: Methuen & Co.
Benedetti, G. (1989). *Psychodynamik der Zwangsneurose*. Darmstadt: Wissenschaftliche Buchgesellschaft.
Bennett, J. M. (1997). Culture Marginality. In M. Paige (Ed.), *Education for Intercultural Experience* (pp. 1–27). Yarmouth, ME: Intercultural Press.
Berrios, G. (1995). Obsessive–Compulsive Disorder. In G. Berrios & R. Porter (Eds.), *A History of Clinical Psychiatry* (pp. 573–592). London: Athlone.
Bien, E. (1927–1928). Die Lehre vom seelischen Zwang im Lichte der Literatur. In W. Stekel, *Zwang und Zweifel*, Vol. I (pp. 557–633). Berlin/Vienna: Urban & Schwarzenberg.
Bien, E. (1933). Der seelische Zwang. In W. Stekel (Ed.), *Der Seelenarzt. Handbuch für seelische Beratung* (pp. 266–306). Leipzig/Amsterdam/Wien: Selbstverlag des Instituts für aktive Psychoanalyse.
Billig, M. (1999). *Freudian Repressions*. Cambridge: Cambridge University Press.
Bleuler, E. (1906). *Affektivität, Suggestibilität, Paranoia*. Halle: Verlag von Carl Marhold.
Bos, J. (1992). On the Origin of the Id (das Es). *The International Review of Psychoanalysis, 4*, 433–443.
Bos, J. (1996). Rereading the Minutes. *Annual of Psychoanalysis, 24*, 223–249.
Bos, J. (2001). Notes on a Controversy: The Question of Lay Analysis. *Psychoanalysis and History, 3*, 153–169.

Bos, J. (2003). A Silent Antipode: The Making and Breaking of Psychoanalyst Wilhelm Stekel. *History of Psychology, 6*, 331–361.
Bos, J. (2004). Everyday Life Objectivity. *Social Epistemology, 2/3*, 123–138.
Bos, J., & L. Groenendijk (2004). The Art of Imitation. *International Journal of Psychoanalysis, 85*, pp. 713–730.
Bos, J., D. Park, & P. Pietikainen (2005). Strategic Self-marginalization: The Case of Psychoanalysis. *Journal of the History of the Behavioral Sciences, 41*, 207–224.
Bourdieu, P. (1988). *Homo Academicus*. Oxford: Polity Press.
Bourdieu, P. (1992). *Language and Symbolic Power*. Oxford: Polity Press.
Bromberg, W. (1963). *The Mind of Man; a History of Psychotherapy and Psychoanalysis* [1954]. New York: Harper Colophon Books.
Brome, V. (1967). *Freud and his Early Circle*. London: Heinemann.
Brown, J. (1967). *Freud and the Post-Freudians*. Harmondsworth: Penguin Books.
Buelens, J. (1971). *Sigmund Freud, kind van zijn tijd; Evolutie en achtergronden van zijn werk tot 1900*. Meppel: Boom.
Burnham, J. C. (1983). *Jelliffe: American Psychoanalyst and Physician & his Correspondence with Sigmund Freud and C. G. Jung*. Chicago: University of Chicago Press.
Caprio, F. C. (1954). *Female Homosexuality*. London: Peter Owen Ltd.
Carrithers, M. (1992). *Why Humans have Cultures. Explaining Anthropology and Social Diversity*. Oxford: Oxford University Press.
Castelnuovo-Tedesco, P. (2002). Stekel, Wilhelm (1868–1940). In E. Erwin (Ed.), *The Freud Encyclopedia. Theory, Therapy, and Culture* (pp. 538–540). New York/London: Routledge.
Christoffel, H. (1939). Abzweigende Richtungen und Schulen. In P. Federn & H. Meng (Eds.), *Das Psychoanalytische Volksbuch* (pp. 667–688). Bern: Hans Huber Verlag.
Clark-Lowes, F. (1997). Wilhelm Stekel's Journalistic Achievement, 1901–1914: The Challenge to Therapeutic Nihilism. *Relation, 4*, 51–74.
Clark-Lowes, F. (2001). Freud, Stekel and the Interpretation of Dreams: The Affinities with Existential Analysis. *Psychoanalysis and History, 3*, 69–78.
Dattner, B. et al. (1965). *Die Onanie; Vierzehn Beiträge zu einer Diskussion der Wiener Psychoanalytischen Vereinigung* (Diskussionen des Wiener psychoanalytischen Vereins, Vol. 2 [1912]). Amsterdam: Bonset.
De Certeau, M. (1988). *The Practice of Everyday Life*. Berkeley: The University of California Press.
Del Pilar, J. A., & J. O. Udasco (2004). Marginality Theory: The Lack of Construct Validity. *Hispanic Journal of Behavioral Sciences, 26*, 3–15.
Douglas, M. (1986). *How Institutions Think*. New York: Syracuse University Press.
Douglas, M. (1999). *Implicit Meanings. Selected Essays in Anthropology*. London/New York: Routledge.
Drinka, G. F. (1984). *The Birth of Neurosis; Myth, Malady, and the Victorians*. New York: Simon & Schuster.
Eissler, K. (1965). *Medical Orthodoxy and the Future of Psychoanalysis*. New York: International Universities Press.
Elias, N. (1956). Problems of Involvement and Detachment. *British Journal of Sociology, 8*, 226–252.
Ellenberger, H. F. (1970). *The Discovery of the Unsconscious. The History and Evolution of Dynamic Psychiatry*. New York: Basic Books.
Fairclough, N. (1992). *Discourse and Social Change*. Cambridge: Polity Press.
Fenichel, O. (1946). *The Psychoanalytic Theory on Neurosis*. London: Routledge & Kegan Paul.
Ferenczi, S., & O. Rank (1922). *Entwicklungsziele der Psychoanalyse*. Vienna: IPV.
Fischer-Homberger, E. (1975). *Die traumatische Neurose. Vom somatischen zum sozialen Leiden*. Bern/Stuttgart/Wien: Hans Huber.
Foucault, M. (1973). *Madness and Civilization*. New York: Vintage Books.
Freeman, W. (1968). *The Psychiatrist; Personalities and Patterns*. New York/London: Grune & Stratton.

Freud, E. (Ed.). (1960). *The Letters of Sigmund Freud*. New York: Basic Books.
Freud, S. (1895a). On the Grounds for Detaching a Particular Syndrome from Neurasthenia under the Description "Anxiety Neurosis". *S.E.* 3.
Freud, S. & J. Breuer. (1895b). Studies on Hysteria. *S.E.* 2.
Freud, S. (1896). The Aethiology of Hysteria. *S.E.* 3.
Freud, S. (1898). Sexuality in the Ethiology of the Neuroses. *S.E.* 3.
Freud, S. (1900). The Interpretation of Dreams. *S.E.* 4/5.
Freud, S. (1901). The Psychopathology of Everyday Life. *S.E.* 6.
Freud, S. (1905). Three Essays on the Theory of Sexuality. *S.E.* 7.
Freud, S. (1907a). Delusions and Dreams in Jensen's *Gradiva*. *S.E.* 9.
Freud, S. (1908a). Hysterical Fantasies and Their Relation to Bisexuality. *S.E.* 9.
Freud, S. (1908b). 'Civilized' Sexual Morality and Modern Nervous Illness. *S.E.* 9.
Freud, S. (1909). Analysis of a Phobia in a Five-Year-Old-Boy. *S.E.* 10.
Freud, S. (1908c). Creative Writers and Day-dreaming. *S.E.* 9.
Freud, S. (1908d). Preface to Wilhelm Stekel's *Nervöse Angstzustände und ihre Behandlung*. *S.E.* 9.
Freud, S. (1910a). Five Lectures on Psycho-analysis. *S.E.* 11.
Freud, S. (1910b). Leonardo da Vinci and a Memory of his Childhood. *S.E.* 11.
Freud, S. (1910c). The Antithetical Meaning of Primal Words. *S.E.* 11.
Freud, S. (1910d). The Future Prospects of Psychoanalytic Therapy. *S.E.* 11.
Freud, S. (1910e). Two Instances of Pathogenic Phantasies Revealed by the Patients Themselves. *S.E.* 11.
Freud, S. (1910f). A Typical Example of a Disguised Oedipus Dream. *S.E.* 5.
Freud, S. (1911). Psycho-analytic Notes on an Autobiographical Case of Paranoia (Dementia Paranoides). *S.E.* 12.
Freud, S. (1912a). Recommendations on Analytic Technique. *S.E.* 12.
Freud, S. (1912b). Contributions to a Discussion on Masturbation. *S.E.* 12.
Freud, S. (1912c). The Dynamics of the Unconscious. *S.E.* 12.
Freud, S. (1912–1913). Totem and Taboo *S.E.* 10.
Freud, S. (1914). On the History of the Psychoanalytic Movement. *S.E.* 14.
Freud, S. (1922). Dreams and Telepathy *S.E.* 18.
Freud, S. (1923). The Ego and the Id. *S.E.* 19.
Freud, S. (1925). An Autobiographical Study. *S.E.* 20.
Freud, S. (1926). The Question of Lay-analysis. *S.E.* 20.
Freud, S. (1930). Civilizations and its Discontents. *S.E.* 21
Freud, S. (1933). New Introductory Lectures on Psycho-analysis. *S.E.* 22.
Freud, S., & K. Abraham (1965). *A Psycho-analytic Dialogue. The Letters of Sigmund Freud and Karl Abraham, 1907–1926*. (B. Marsch & H. C. Abraham, Eds.). New York: Basic Books.
Freud, S., & L. Andreas-Salomé (1972). *Letters*. (E. Pfeiffer, Ed.). New York: W.W. Norton.
Freud, S., & L. Binswanger (1992). *Briefwechsel 1908–1938*. (G. Fichtner, Ed.). Frankfurt: Fischer Verlag.
Freud, S., & S. Ferenczi (1993). *Briefwechsel Bd. I/1 and I/2 (1908–1914)*. Vienna: Böhlau Verlag.
Freud, S., & G. Groddeck (1974). *Briefe über das Es*. München: Kindler Verlag.
Freud, S., & E. Jones (1993). *The Complete Correspondence of Sigmund Freud and Ernest Jones*. (R. A. Paskauskas, Ed.). London: Harvard University Press.
Freud, S., & C. G. Jung (1974). *The Freud/Jung Letters*. (W. McGuire, Ed.). London: Hogarth Press and Routledge.
Gay, P. (1984). *The Bourgeois Experience. Victoria to Freud*. Vol. I. *Education of the Senses*. New York: Oxford University Press.
Gay, P. (1988). *Freud: A Life for our Time*. London: J.M. Dent.
Gay, P. (1995). *The Bourgeois Experience. Victoria to Freud*. Vol. 4. *The Naked Heart*. New York: W.W. Norton & Co.
Gilman, S. L. (1993). *The Case of Sigmund Freud: Medicine and Identity at the Fin de siècle*. Baltimore: Johns Hopkins University Press.

Goethe, J. W. von (1995). *Wilhelm Meister's Apprenticeship*. In E. Blackall (Ed.), *The Collected Works*. Vol. 9. New Jersey: Princeton University Press.

Goffman, E. (1963). *Stigma. Notes on the Management of Spoiled Identity*. New York: Prentice-Hall.

Grinstein, A. (1958). *The Index of Psychoanalytic Writings* (Vol. IV). New York: International University Press.

Groenendijk, L. F. (1996). Die kulturkritische Familienpädagogik des Psychoanalytikers Wilhelm Stekels. In J. Sturm, J. Dekker, R. Aldrich, & F. Simon (Eds.), *Education and Cultural Transmission* [Paedagogica Historica, Suppl. Series, Vol. II] (pp. 145–162). Gent: CSHP.

Groenendijk, L. (1997). Masturbation and Neurasthenia: Freud and Stekel in Debate on the Harmful Effects of Autoerotism. *Journal of Psychology and Human Sexuality, 9*, 71–94.

Groenendijk, L. (2001). Aufklärung statt Prügel. Der Psychoanalytiker Wilhelm Stekel (1868–1940) über Ursachen und Überwindung der "Erziehung zur Grausamkeit." In V. Schmid (Ed.), *Verwahrlosung—Devianz—Antisoziale Tendenz. Stränge zwischen Sozial-und Sonderpädagogik* (pp. 146–240). Freiburg: Lambertus.

Groenendijk, L., & J. Sturm (2000). Curing Society by Better Education. Psychoanalyst Wilhelm Stekel (1868–1940) on Parental Influences in the Causation and Prevention of Compulsion Neuroses. *Peadagogica Historica, 36*, 635–673.

Groenendijk, L. F., & N. Bakker (2002). Child Rearing and the Neuroticization of Parenting: The Case of the Netherlands. *History of Education, 31*, 571–588.

Gross, M. L. (1979). *The Psychological Society. A Critical Analysis of Psychiatry, Psychotherapy, Psychoanalysis and the Psychological Revolution*. New York: Simon & Schuster.

Grosskurth, Ph. (1991). *The Secret Ring; Freud's Inner Circle and the Politics of Psychoanalysis*. Reading, MA: Addison-Wesley.

Grünbaum, A. (1984). *The Foundations of Psychoanalysis. A Philosophical Critique*. Berkeley: The University of California Press.

Gutheil, E. A. (1962). "Preface". In W. Stekel, *Compulsion and Doubt* (pp. 1–31.) New York: Grosset & Dunlop.

Hale, N. G. (1995). *The Rise and Crisis of Psychoanalysis in the United States. Freud and the Americans, 1917–1985*. New York/Oxford: Oxford University Press.

Handlbauer, B. (1990). *Die Adler-Freud-Kontroverse*. Munich: Fischer Verlag.

Harper, R. A. (1959). *Psychoanalysis and Psychotherapy. 36 Systems*. Englewood Cliffs, NJ: Prentice-Hall.

Harris, J. R. (1998). *The Nurture Assumption. Why Children Turn Out the Way They Do*. New York: The Free Press.

Horney, K. (1966). *New Ways in Psychoanalysis*. New York: W.W. Norton & Comp.

Israëls, H., & M. Schatzman (1993). The Seduction Theory. *History of Psychiatry, 4*, 23–59.

Jackson, S. W. (1986). *Melancholia and Depression; From Hippocratic Times to Modern Times*. New Haven/London: Yale University Press.

Jaspers, K. (1948). *Allgemeine Psychopathologie* [General psychopathology]. Berlin: Springer Verlag.

Jensen, W. (1929). Drei unveröffentlichte Briefe. Zur Geschichte von Freuds Gradiva-Analyse [Three unpublished Letters. On the History of Freud's Gradiva Study]. *Die Psychoanalytische Bewegung, 1*, 207–211.

Jerábek, R. (1994). Wilhelm Stekel. In O. Frischenschlager (Ed.), *Wien, wo sonst!; Die Entstehung der Psychoanalyse und ihrer Schulen*. Wien/Köln/Weimar: Böhlau.

Johnson, S. M. (1994). *Character Styles*. New York/London: Norton & Comp.

Jones, E. (1953–1957). *The Life and Work of Sigmund Freud* (3 Vols). New York: Basic Books.

Jones, E. (1959). *Free Associations*. New York: Basic Books.

Jung, C. G. (1902). *Zur Psychologie und Pathologie sogenannter occulter Phänomene* [On the Psychology and Pathology of so-called Occult Phenomena]. Leipzig: Oswald Mutze.

Jung, C. G. (1912). *Wandlung und Symbole der Libido*. Leipzig/Vienna: Deuticke.

Kerckhoff, A. C., & T. C. McCormick (1955). Marginal Status and Marginal Personality. *Social Force, 1*, 48–55.

Kerr, J. (1994). *A Most Dangerous Method: The Story of Jung, Freud & Sabina Spielrein*. New York: Vintage Books.

Kretschmer, E. (1966). *Der sensitive Beziehungswahn; Ein Beitrag zur Paranoiafrage und zur psychiatrischen Charakterlehre*. Berlin/Heidelberg/New York: Springer.

Krüll, M. (1986). *Freud and his Father*. New York/London: Norton.

Kuhn, P. (1998). "A Pretty Piece of Treachery": The Strange Case of Dr. Stekel and Sigmund Freud. *International Journal of Psycho-Analysis, 79*, 1151–1171.

Laqueur, T. W. (2003). *Solitary Sex: A Cultural History of Masturbation*. New York: Zone Books.

Levenson, E. (1983). *The Ambiguity of Change. An Inquiry into the Nature of Psychoanalytic Reality*. New York: Basic Books.

Leupold-Löwenthal, H. (1980). The Minutes of the Vienna Psycho-Analytic Society. *Sigmund Freud House Bulletin, 2*, 23–41.

London, L. S., & F. S. Caprio (1950). *Sexual Deviations*. Washington: The Linacre Press.

Lumsden, D. (1984). *Community Mental Health in Action*. Toronto: Canadian Mental Health Association.

Lütkehaus, L. (1992). *"O Wollust, o Hölle"; Die Onanie: Stationen einer Inquisition*. Frankfurt am Main: Fischer Taschenbuch Verlag.

Mahony, P. (1987). *Freud as a Writer*. New Haven, CT: Yale University Press.

Masson, J. M. (1985). *The Assault on Truth. Freud's Suppression of the Seduction Theory*. New York: Penguin.

Mayo, E. (1951). *The Psychology of Pierre Janet*. London: Routledge & Kegan Paul.

McLaughlin, N. (2001). Optimal Marginality. *The Sociological Quarterly, 42*, 271–288.

Merton, R. (1965). *On the Shoulders of Giants. A Shandean Postscript*. New York: The Free Press.

Mitchell, J. (1979). *Psychoanalysis and Feminism*. Harmondsworth: Penguin Books.

Moore, B. E., & B. D. Fine (Eds.). (1990). *Psychoanalytic Terms and Concepts*. New Haven/London: The A.P.A. & Yale U.P.

Mortier, F., W. Colen & F. Simon (1994). Inner-scientific Reconstructions in the Discourse on Masturbation (1760–1950). *Paedagogica Historica, 30*, 817–847.

Mühlleitner, E., & J. Reichmayr (1992). *Biographisches Lexikon der Psychoanalyse: Die Mitglieder der Psychologischen Mittwoch-Gesellschaft und der Wiener Psychoanalytischen Vereinigung, 1902–1938*. Tübingen: edition diskord.

Nagera, H. (1991). *Psychoanalytische Grundbegriffe; Eine Einführung in Sigmund Freuds Terminologie und Theoriebildung*. Frankfurt am Main: Fischer Taschenbuch Verlag.

Nitzschke, B. (1992). Wilhelm Stekel, ein Pionier der Psychoanalyse—Anmerkungen zu ausgewählten Aspekten. In E. Federn & G. Wittenberger (Eds.), *Aus dem Kreis um Sigmund Freud* (pp. 176–191). Frankfurt: Fischer Verlag.

Nunberg, H., & E. Federn (1962–1975). *Minutes of the Vienna Psychoanalytic Society* (4 Vols). New York: International Universities Press.

Nunberg, H., & E. Federn (1976–1981). *Protokolle der Wiener Psychoanalytischen Vereinigung* (4 Vols). Frankfurt am Main: Fischer.

Oerlemans, A. C. (1949). *Development of Freud's Conception of Anxiety*. Amsterdam: Noord-Hollandsche Uitgeversmaatschappij.

Park, D. W. (2004). The Couch and the Clinic: The Cultural Authority of Popular Psychiatry and Psychoanalysis. *Cultural Studies, 18*, 109–133.

Placzek, R. (Ed.). (1981). *Record of a Friendship. The Correspondence of Wilhelm Reich and A. S. Neill*. New York: Farrer, Strauss and Giroux.

Pohlen, M., & M. Bautz-Holzherr (1995). *Psychoanalyse – das Ende einer Deutungsmacht*. Reinbek bei Hamburg: Rowohlt Taschenbuch Verlag.

Rachman, S. J., & R. J. Hodgson (1980). *Obsessions and Compulsions*. Englewood Cliffs, NJ: Prentice-Hall.

Rattner, J. (1990). *Klassiker der Tiefenpsychologie*. München: Psychologie Verlags Union.

Rapoport, J. L. (1989). *The Boy who Couldn't Stop Washing*. New York: E.P. Dutton.

Rehm, W. (1968). *Die psychoanalytische Erziehungslehre; Anfänge und Entwicklung*. München: Piper & Co.

Reich, A. (1951). The Discussion of 1912 on Masturbation and our Present-day views. *The Psychoanalytic Study of the Child, 6*, 80–94.

Roazen, P. (1975). *Freud and his Followers*. New York: Meridan Books.

Roazen, P. (1986). *Brother Animal. The Story of Freud and Tausk*. New York: New York University Press.

Roazen, P. (2001). *The Historiography of Psychoanalysis*. New Brunswick: Transaction Publishers.

Ronge, P. H. (1934). *Individualpsychologie. Een systematische uiteenzetting*. Utrecht: Bijleveld.

Roudinesco, E., & M. Plon (1997). *Dictionnaire de la psychanalyse*. Paris: Fayard.

Rudnytsky, P. (2002). *Reading Psychoanalysis. Freud, Rank, Ferenczi, Groddeck*. Ithaca: Cornell University Press.

Sachs, H. (1945). *Freud: Master and Friend*. London: Imago.

Sadger, I. (2005). *Recollecting Freud* (edited and introduced by A. Dundes). Madison: The University of Wisconsin Press.

Salzman, L. (1980). *Treatment of the Obsessive Personality*. New York/London: Jason Aronson.

Schafer, R. (1992). Reading Freud's Legacies. In J. H. Smith & M. Morris (Eds.), *Telling Facts: History and Narration in Psychoanalysis* (pp. 1–20). Baltimore/London: The Johns Hopkins University Press.

Schindler, W. (1980). *Aktive Psychoanalyse Eklektisch Gesehen. Ein Lesebuch zusammengestellt, kommentiert, mit eigenen Fällen ergänzt und herausgegeben*. Bern: Hans Huber Verlag.

Schneider, K. (1940). *Die psychopathischen Persönlichkeiten* [Orig. 1923]. Vienna: F. Deuticke.

Shorter, E. (1992). *From Paralysis to Fatigue; a History of Psychosomatic Illness in the Modern Era*. New York: Free Press.

Showalter, E. (1987). *The Female Malady. Women, Madness, and English Culture, 1830–1980*. London: Virago Press.

Solomon, I. (1969). Book Review of "The Minutes of the Vienna Psychoanalytic Society," Vol. II. *Psychoanalytic Quarterly, 38*, 473–478.

Stanton, M. (1988). Wilhelm Stekel: A Refugee Analyst and his English Reception. In E. Timms & N. Segal (Eds.), *Freud in Exile. Psychoanalysis and its Vicissitudes* (pp. 163–174). New Haven: Yale University Press.

Stanton, M. (1991). *Sándor Ferenczi: Reconsidering Active Intervention*. Northvale, NJ/London: Jason Aronson.

Stanton, M. (1992). The Case of Otto Gross: Jung, Stekel, and the Pathologization of Protest. In E. Timms & R. Robertson (Eds.), *Austrian Studies III: Psychoanalysis in its Cultural Context* (pp. 49–56). Edinburgh: Edinburgh University Press.

Stekel, W. (1895). Ueber Coitus im Kindesalter [On Coitus in Childhood]. *Wiener Medizinische Blätter, 18*, 247–249.

Stekel, W. (1902 [1986]). Traumleben und Traumdeutung (Review of Freud's *Traumdeutung*). In G. Kimmerle (Ed.), *Freud's Traumdeutung, Frühe Rezensionen 1899–1903*. Tübingen: edition diskord.

Stekel, W. (1903 [1986]). Der "Kleine Kohn" [Little Kohn]. In E. Federn & G. Wittenberger (Eds.), *Aus dem Kreis um Sigmund Freud* (pp. 220–227). Frankfurt: Fischer Verlag.

Stekel, W. (1907a). *Die Ursachen der Nervosität. Neue Ansichten über deren Entstehung und Verhütung* [Causes of Nervousness. New Viewpoints on its Cause and Prevention]. Vienna: Knepler Verlag.

Stekel, W. (1907b). Nervöse Angstzustände und ihre Behandlung [Nervous Anxiety States and their Treatment]. *Medizinische Klinik, 3*, 1039–1040, 1064–1067.

Stekel, W. (1908). *Nervöse Angstzustände und ihre Behandlung* [Nervous Anxiety States and their Treatment]. Berlin/Vienna: Urban & Schwarzenberg.

Stekel, W. (1909a). Beiträge zur Traumdeutung [Contributions to Dream Interpretation]. *Jahrbuch für psychoanalytische und psychopathologische Forschungen, 1*, 458–512.

Stekel, W. (1909b). *Dichtung und Neurose. Bausteine zur Psychologie des Künstlers und des Kunstwerkes* [Poetry and Neurosis. Building Blocks for a Psychology of Art and Artists]. Wiesbaden: Bergmann.
Stekel, W. (1910). Die Frage der Abstinenz [The Problem of Abstinence]. *Zentralblatt für Psychoanalyse, 1*, 68–69.
Serenus [ps. W. Stekel]. (1911a). *Äskulap als Harlekin. Humor, Satire und Phantasie aus der Praxis* [Aesculapius as Harlequin. Humor, Satire and Fantasy from a Medical Practice]. Wiesbaden: Bergmann.
Stekel, W. (1911b). *Nervöse Leute* [Nervous People]. Vienna: Paul Knepler.
Stekel, W. (1911c). *Die Sprache des Traumes. Eine Darstellung der Symbolik und Deutung des Traumes in ihren Beziehungen zur kranken und gesunden Seele* [The Language of Dreams. An Exposition on Symbolism and Interpretation of Dreams in Relation to Mental Health and Illness]. Munich/Vienna: Bergmann.
Stekel, W. (1912a). Der Zweifel. [Doubt] *Zeitschrift für Psychotherapie und medizinische Psychologie, 4*, 332–355.
Stekel, W. (1912b). *Die Träume der Dichter. Eine vergleichende Untersuchung der unbewussten Triebkräfte bei Dichtern, Neurotikern und Verbrechern. Bausteine zur Psychologie des Klinikers und des Kunstwerkes* [The Dreams of Poets. A Comparative Analysis of Unconscious Drives in Poets, Neurotics and Criminals]. Wiesbaden: Bergmann.
Stekel, W. (1912–1928). *Störungen des Trieb-und Affektlebens* (10 Vols) [Disorders of the Instincts and the Emotions]. Berlin/Vienna: Urban & Schwarzenberg.
Stekel, W. (1920a). *Die impotenz des Mannes* [Impotence in the Male]. Vienna: Urban & Schwarzenberg.
Stekel, W (1921a). *Nervöse Angstzustände und ihre Behandlung* (3rd ed.) [Nervous Anxiety States and their Treatment]. Berlin/Vienna: Urban & Schwarzenberg.
Stekel, W. (1921b). *Die Geschlechtskälte der Frau, eine Psychopathologie des weiblichen Liebeslebens.* [Frigidity in Woman in Relation to her Love Life]. Berlin/Vienna: Urban & Schwarzenberg.
Stekel, W. (1922a). *Die Sprache des Traumes. Eine Darstellung der Symbolik und Deutung des Traumes in ihren Beziehungen zur kranken und gesunden Seele* (3rd ed.) [The Language of Dreams. An Exposition on Symbolism and Interpretation of Dreams in Relation to Mental Health and Illness]. Wiesbaden: Bergmann.
Stekel, W. (1922b). *Psychosexueller Infantilismus. Die seelischen Kinderkrankheiten der Erwachsenen* [Psychosexual Infantilism. Psychical Illnesses of Adults]. Berlin/Vienna: Urban & Schwarzenberg.
Stekel, W. (1923a). Alfred Adler, Panacea-ist. *Modern Quarterly*, 524–527.
Stekel, W. (1923b). *Conditions of Nervous Anxiety and their Treatment*. London: Kegan Paul, Trench, Trubner & Co.
Stekel, W. (1923c). *Onanie und Homosexualität (Die homosexuelle Neurose)* [Masturbation and Homosexuality (The Homosexual Neurosis)] (3rd ed.). Berlin/Vienna: Urban & Schwarzenberg.
Stekel, W. (1924). *Masken der Sexualität (der innere Mensch)* [Masks of Sexuality]. Vienna: Knepler Verlag.
Stekel, W. (1926). Zur Geschichte der analytischen Bewegung [On the History of the Analytical Movement]. *Fortschritte der Sexualwissenschaft und Psychoanalyse, 2*, 539–575.
Stekel, W. (1927–1928). *Zwang und Zweifel. Für Ärzte und Mediziner dargestellt* (2 Vols) [Compulsion and Doubt. An Exposition for GP's and Medical Doctors]. Berlin/Vienna: Urban & Schwarzenberg.
Stekel, W. (1927–1929). *Briefe an eine Mutter*, Vols. I–III [Letters to a mother]. Zürich/Leipzig: Wendepunkt Verlag.
Stekel, W. (1931). *Marriage at the Crossroads*. New York: William Godwin.
Stekel, W. (Ed.). (1933). *Der Seelenarzt; Handbuch für seelische Beratung* [The Soul Doctor. Handbook for Psychical Healing]. Leipzig/Amsterdam/Wien: Selbstverlag des Instituts für Aktive Psychoanalyse.

Stekel, W. (1934). *Erziehung der Eltern* [Education of the Parents]. Vienna: Verlag der psychotherapeutischen praxis Weidmann & Co.
Stekel, W. (1935). *Fortschritte und Technik der Traumdeutung* [The Interpretation of Dreams]. Vienna: Verlag für Medizin Weidmann & Co.
Stekel, W. (1937). l'Évolution, les méthodes et les dangers de la psychanalyse. In W. Stekel, *La Femme Frigide* (pp. 11–38). Paris: Gallimard.
Stekel, W. (1950a). *The Autobiography of Wilhelm Stekel.* (E. A. Gutheil, Ed.). New York: Liveright.
Stekel, W. (1950b). *Technique of Analytical Psychotherapy* [translation of *Technik der analytischen Psychotherapie*]. New York: Liveright.
Stekel, W. (1950c). *Auto-erotism: A Psychiatric Study of Masturbation and Neurosis.* New York: Liveright.
Stekel, W. (1962a). *Compulsion and Doubt* [Zwang und Zweifel]. New York: Grosset & Dunlap.
Stekel, W. (1962b). *The Interpretation of Dreams* [translation of *Fortschritte und Technik der Traumdeutung*]. New York: Grosset & Dunlap.
Stekel, W. (1966). *Patterns of Psychosexual Infantilism* [translation of *Psychosexueller Infantilismus*]. New York: Washington Square Press.
Stekel, W. (1968). *Sadism and Masochism. The Psychology of Hatred and Cruelty* [translation of *Sadismus und Masochismus*]. New York: Washington Square Press.
Stekel, W. (1971). *Impotence in the Male* (2 Vols) [translation of *Die Impotenz des Mannes*]. New York: Liveright.
Stone, M. H. (1997). *Healing the Mind. A History of Psychiatry from Antiquity to the Present.* New York: Norton & Comp.
Stonequist, V. E. (1937). *The Marginal Man*. New York: Scribner's.
Sulloway, F. J. (1979). *Freud, Biologist of the Mind.* Bungay: Fontana Books.
Szasz, T. S. (1970). *The Manufacture of Madness.* New York: Harper & Row.
Tallis, F. (1995). *Obsessive Compulsive Disorder. A Cognitive and Neuropsychological Perspective.* Chichester: John Wiley & Sons.
Thompson, J. B. (1995). *The Media and Modernity: A Social Theory of the Media.* Stanford, CA: Stanford University Press.
Timms, E. (1986). *Karl Kraus: Apocalyptic Satirist.* New Haven, CT: Yale University Press.
Timms, E. (1992). Introductory Note to "From the Memoirs of a Freudian". In E. Timms & R. Roberston (Eds.), *Psychoanalysis. Austrian Studies* (Vol. III). Edinburgh: Edinburgh University Press.
Timms, E. (1995). *Freud and the Child Woman. The Memoirs of Fritz Wittels.* New Haven: Yale University Press.
Timms, E. (1996). In Freud's shadow. The Double Exile of Wilhelm Stekel. In C. Brinson, R. Dove, M. Malet, & J. Taylor (Eds.), *"England? Aber wo liegt es?" Deutsche und österreichische Emigranten in Grossbritanien 1933–1945* ["England? Where is that?" German and Austrian Immigrants in Great Britain 1933–1945] (pp. 47–58). München: Iudicium Verlag.
Walker Puner, H. (1947). *Sigmund Freud: his Life and Mind.* New York: Dell Publishing Co.
Wallerstein, R. (1998). *Lay Analysis. Life Inside the Controversy.* London: The Analytic Press.
Way, L. (1950). *Adler's Place in Psychology.* New York: Macmillan.
Wexberg, E. (1931). *Individualpsychologie. Eine systematische Darstellung.* Leipzig: Hirzel.
Wittels, F. (1924). *Sigmund Freud: His Personality, his Teachings and his School.* London: George Allen & Unwin.
Wittenberger, G., & H. Tögel (Eds.). (1999–2004). *Die Rundbriefe des geheimen Kommittees* (4 Vols). Tübingen: edition diskord.
Wolman, B. B., & G. Stricker (Eds.). (1994). *Anxiety and Related Disorders. A Handbook.* New York: John Wiley & Sons.
Wolff, H. K. (1966). *Compulsive Desire.* Los Angeles: Holloway House Publishing.
Wortis, J. (1954). *Fragments of an Analysis With Freud.* New York: Simon and Schuster.

INDEX

9 780387 326993

MIX
Papier aus verantwortungsvollen Quellen
Paper from responsible sources
FSC® C105338

If you have any concerns about our products,
you can contact us on
ProductSafety@springernature.com

In case Publisher is established outside the EU,
the EU authorized representative is:
Springer Nature Customer Service Center GmbH
Europaplatz 3, 69115 Heidelberg, Germany

Printed by Libri Plureos GmbH
in Hamburg, Germany